Jae-Ho Hwang

Eco-friendly utilization of marine fish skin in Korea

AF154053

Jae-Ho Hwang

Eco-friendly utilization of marine fish skin in Korea

LAP LAMBERT Academic Publishing

Impressum / Imprint

Bibliografische Information der Deutschen Nationalbibliothek: Die Deutsche Nationalbibliothek verzeichnet diese Publikation in der Deutschen Nationalbibliografie; detaillierte bibliografische Daten sind im Internet über http://dnb.d-nb.de abrufbar.
Alle in diesem Buch genannten Marken und Produktnamen unterliegen warenzeichen-, marken- oder patentrechtlichem Schutz bzw. sind Warenzeichen oder eingetragene Warenzeichen der jeweiligen Inhaber. Die Wiedergabe von Marken, Produktnamen, Gebrauchsnamen, Handelsnamen, Warenbezeichnungen u.s.w. in diesem Werk berechtigt auch ohne besondere Kennzeichnung nicht zu der Annahme, dass solche Namen im Sinne der Warenzeichen- und Markenschutzgesetzgebung als frei zu betrachten wären und daher von jedermann benutzt werden dürften.

Bibliographic information published by the Deutsche Nationalbibliothek: The Deutsche Nationalbibliothek lists this publication in the Deutsche Nationalbibliografie; detailed bibliographic data are available in the Internet at http://dnb.d-nb.de.
Any brand names and product names mentioned in this book are subject to trademark, brand or patent protection and are trademarks or registered trademarks of their respective holders. The use of brand names, product names, common names, trade names, product descriptions etc. even without a particular marking in this works is in no way to be construed to mean that such names may be regarded as unrestricted in respect of trademark and brand protection legislation and could thus be used by anyone.

Coverbild / Cover image: www.ingimage.com

Verlag / Publisher:
LAP LAMBERT Academic Publishing
ist ein Imprint der / is a trademark of
OmniScriptum GmbH & Co. KG
Heinrich-Böcking-Str. 6-8, 66121 Saarbrücken, Deutschland / Germany
Email: info@lap-publishing.com

Herstellung: siehe letzte Seite /
Printed at: see last page
ISBN: 978-3-659-59410-6

Contents

Introduction

To minimize the resources wasted in the world, developing a new value creation and growth power has been magnified as the main factors influencing the competitiveness of the national economy. Accordingly, the Korean President Park Geun-Hye has been accelerating the realization of a resource recycling society on resources and energy as national issues from the initial launch. In order to achieve early successful establishment of these national issues, the government legislates "resource circulation society conversion promotion" by 2014, and plans to introduce a "waste reclamation and incineration disposal burden" which exceeds the cost of reclamation and incineration on waste resource rather than cost of resource circulation recycling. An urgent response is required fisheries strategy in rushing the transition to future resource recycling society at the national level in order to minimize such impact on fisheries and to respond actively to the national philosophy.

Fishery products are easy to damage its biological nature, quality, or to decompose. There is a significant amount in the inedible parts of fishery products such as bone, fin, and intestine of fish and shells of shellfish. Substantial volume of by-product is come out at the stages of production, distribution, processing, and sales. In the case of simple disposal on fisheries by-products, environment pollution, landscape damage, and odor-causing as well as increase of disposal costs induce in the procedures of production, distribution, processing, and sales. However, in the case of utilizing the fisheries by-products by eco-friendly or industrially, it is possible to develop the new value creation and the new growth power, as well as solving the above problems at the same time. Focusing on this point, it has been constantly trying to convert simply discarded fishery by-products to eco-friendly resources or industrially high value resources in the world.

Accordingly, here we introduce the 3 latest published papers. First, Korean four marine fish skins are characterized by biochemical techniques. Second, dietary flounder skin meal, which is superior to the 4 skins, is added to substitute the feed protein source, and the feeding trial improves growth performance, body composition, and stress recovery in the juvenile black rockfish. Third, edible parts of the cultured fish are analyzed by nutritional characteristics.

Chapter 1

Biochemical characteristics of four marine fish skins in Korea

In this study, we investigated the biochemical characteristics of the fish skins of 4 industrial species, olive flounder (*Paralichthys olivaceus*), black rockfish (*Sebastes schlegeli*), sea bass (*Lateolabrax maculatus*), and red sea bream (*Pagrus major*).   There is high domestic demand in Korea for farming of these fish for human consumption.   Crude protein contents in the skin of these fish ranged from 73% to 94% by dry weight; this was in part due to a high content of structural protein, collagen.   Among the 4 species, olive flounder had the thickest dermal and epidermal layers in the dorsal skin.   This species was also associated with the highest extraction ratio of acid-soluble collagen.   We also examined whether fish skin could be a cost-effective alternative to current fish meal sources.   Our analysis indicates that, when supplemented with additional fish oils and essential amino acids, fish skin is a viable alternative for fish meal formulations.

Introduction

Improved fishing equipment, fishing techniques and increased aquaculture production has lead to an increase in Korean domestic seafood production from 470,000 tons in the early 1960s to 3 million tons in the early 1990s.   Production has since remained constant, and ranges between 2.5 and 3 million tons annually.   In particular, fish comprised more than 40% of the total seafood produced from 2007 to 2009, and farmed marine fishes accounted for approximately 7% of the total fish production.   Although farmed fish represent only a small fraction of total fish production, they are one of the major sources of sashimi and sushi products (Park & Kim, 2007).   Wild-caught fish used in dried fish production are mainly

2

anchovy and saurel or large fish such as king mackerel, tuna, and yellowtail, which are generally used for broiling (2010 Fishery Production Statistics Yearbook).   Olive flounder, black rockfish, sea bass, and red sea bream are major marine fish species farm-raised in Korea, accounting for over 90% of the total aquaculture production.   Although these species are also wild-caught because of domestic consumers' preference for wild fish, catch-dependent price fluctuations and small production volumes ensures a continued consumption of the more cost-effective farm-raised fish.

As the muscle tissue of most farmed sea fish is consumed for sashimi, large quantities of the remnants, including head, bones, guts, and skin are left as by-products, which are usually discarded.   Olive flounder, black rockfish, sea bass, and red sea bream account for the major part of this waste, as they are the most popular fish that are consumed domestically. Although the head parts, bones, and guts can be made into a stew that is served as the last course of the meal, the skin is not used, since it is removed at the initial step of preparation and discarded.

Fish skin is mainly composed of high molecular weight elastic protein and has a high concentration of connective tissue protein, collagen (Jayathilakan et al., 2012).   While studies on the extraction and use of collagen and gelatin derived from fish skin have been performed (Yoo et al., 2008), no studies have explored the possibility that fish skin could be exploited as a protein source.   With this in mind, we designed the present study to describe the biochemical and histological characteristics of the skins of olive flounder, black rockfish, sea bass, and red sea bream, as well as the chemical characteristics of acid-soluble collagen (ASC) extracted from these skins.   Furthermore, the feasibility of using fish skin as a protein source was evaluated by conducting a sodium dodecyl sulfate polyacrylamide gel electrophoresis (SDS-PAGE)-based comparative analysis of fish skin meal.

3

Material and Methods

Fish skin procurement

The fish skins of olive flounder, black rockfish, sea bass, and red sea bream were collected from the local fish market in Namsan-dong, Yeosu-si, Korea. The collected skins, which had been discarded after sashimi preparation, were stored in ice. Skin was cleared of any residual muscle, and washed, after which the skins were separated according to the species and kept at -45°C in a freezer (SW-UF-700P; Samwon, Korea).

Analysis of basic components

The moisture level of frozen fish skin was measured with an atmospheric-pressure heat-drying type moisture gauge (HR 73 Halogen Moisture Analyzer, Switzerland) immediately after being cut into 1-cm^2 pieces. The fish skin pieces were dried in a dry oven (Jesco, Japan) at 60°C for 48 h. They were then homogenized with a high-speed grinder (High Speed ZM-1000, Retch Co., Japan) and subjected to basic component analyses according to the methods established by AOAC method (2002). In detail, crude protein was analyzed with an automatic analyzer (KJELTEK Auto Sampler System 1035, Switzerland) based on the Kjeldahl nitrogen determination method ($N \times 6.25$), crude fat was analyzed with an automatic extractor (Soxtec 2050 Auto Extraction Unit, Switzerland) using an ether extraction method, and crude ash was analyzed using a direct ashing method (Electric furnace, TMF-3100; EYELA, Japan) by burning at 550°C.

Fatty acyls

After methyl esterization of fatty acyls according to the AOCS method (1990), we performed gas-liquid chromatography (Shimadzu GC 17A, Shimadzu Seisakusho Co. Ltd., Kyoto, Japan) with a built-in capillary column (SP^{TM}-2560 capillary column; 100 m length $\times$ 0.25 mm internal diameter $\times$ 0.25-μm film thickness). During pretreatment, 20 mL methanol and 10 mL chloroform was added to 2 g of each sample in a 100-ml triangular flask, and samples were homogenized for 10 min.   Following the addition of 10 mL chloroform, a further homogenization was then performed.   After the homogenized liquid was filtered (1-μm filter) and concentrated under reduced pressure, fatty acyls (250 μL) dissolved in 1 mL toluene were methylated for 30 min in 750 μL of trimethylsulfonium hydroxide. One milliliter of methylated fatty acyls was harvested in a vial and analyzed with gas-liquid chromatography (Shimadzu GC 17A; Shimadzu Seisakusho Co. Ltd., Kyoto, Japan).   For analysis, the injector (Fid) temperature was set at 250°C, and the column temperature was kept at 180°C for 8 min, raised to 230°C at 3°C/min, and maintained for 15 min.   The carriage gas was helium (transfer fluid pressure, 1.0 kg/cm^2) with a split ratio of 1:50.   The internal standard was methyltricosanoate (Aldrich Chem. Co., Milwaukee, WI, USA).

Amino acid composition

For each species, 0.5 g of fish skin and meal was weighed in an 18-mL test tube and added to 3 mL of 6N HCl.   The test tube was sealed with a vacuum pump and maintained in a heating block for hydrolysis at 121°C for 24 h.   After the removal of acids with a rotary evaporator (50°C, 40 psi), 10 mL of the solution was weighed with sodium loading buffer, of which 1 mL was filtered through a membrane filter (0.2 μL) for quantitative analysis using an amino-acid analyzer (S-433H; SYKAM GmbH, Germany).   The analytic conditions were as follows: cation separation column (LCA K06/Na); column size, 4.6 × 150 mm; column temperature, 57°C~74°C; buffer flow rate, 0.45 mL/min; reagent flow rate, 0.25 mL/min; buffer pH range, 3.45–10.85; and wavelengths, 440 nm and 570 nm.

Histological analysis

We examined the histological characteristics of live farm-raised olive flounder, black rockfish, sea bass, and red sea bream that were purchased at the local fish market in Namsan-dong, Yeosu-si, Jeollanam-do (south-western province of South Korea).   The fish were

5

anesthetized with 100 ppm fish sedative (AQUIS-S, New Zealand), and the dorsal and ventral parts including skin and muscle tissues were excised for histological analyses.

Fish skin was cut into 1-cm^2 pieces for a histological examination that was performed with an optical microscope. The sample pieces were fixed in aqueous Bouin's fluid (75 mL saturated picric acid solution, 25 mL formaldehyde, 5 mL acetic acid) for 12–24 h, according to the method of Drury and Wallington (1980), and then washed in running water for 24–48 h. Graded ethanol dehydration (70%-100%, 40 min per stage) was then performed, and this was followed by replacement of the alcohol using xylene (xylene I, II, III; 40 min each) and paraplast (McCormick, USA) embedding. The embedded samples were sliced into 4- to 6-µm-thick sections using a microtome (RM2235; Leica, Germany) and were attached to glass slides.

Mayer's hematoxylin-eosin (H&E) staining was performed to observe tissue architecture and cell types. Masson's trichrome staining identified the composition of tissues and cells. Alcian blue-periodic acid and Schiff's solution (AB-PAS, pH 2.5) were used to identify the mucopolysaccharide component. After staining, the sections were subjected to phase dehydration and a clearing process using xylene, and then sealed in Canada balsam.

After the paraffin-free tissues were washed step by step in 100, 90, 80, and 70% alcohol, Masson's trichrome staining was performed in the order of Weigert's iron hematoxylin staining (10 min), washing (20-30 min), Biebrich scarlet-acid fuchsin staining (15 min), phosphomolybdic acid-phosphotungstic acid (10–15 min), aniline blue (5–10 min), and 1% acetic water (3–5 min).

For the AB-PAS (pH 2.5) reaction, sections were washed in 90% alcohol after removing paraffin with xylene. The sections were then incubated with periodic acid, Schiff's solution, and sodium sulfite water I, II and III for 5, 10, and 10 min each, respectively, before a final 5-min stain with Mayer's hematoxylin.

The samples were observed with an optical microscope (CX31; Olympus, Japan), and the skin thickness of each fish species was measured using the Image Measurement System (Focus

Technology).   The average thickness was calculated from 30 repeated measurements of 20 slides for each species.

Fish skin collagen ratio and molecular characteristics
ASC extraction

ASC concentrations in the skins of the 4 species were measured according to the methods of Yoshinaka et al (1985) and Hwang et al (2007).   In order to avoid the influence of proteases and to eliminate non-collagen protein, the fish skin stored at -45°C was diced into approximately 1-cm^2 sections at low temperature (5°C). Samples were homogenized with 0.1 M NaOH [20 times the original sample quantity (v/w)], stirred for 24 h, and cleared from the supernatant by centrifugation (4°C, 10,000 × g, 20 min).   This process was repeated 5 times, followed by 3 washes with cold distilled water, after which the residue after alkali extraction (RS-AL) was obtained.   The RS-AL was stirred for 24 h after the addition of 0.5 M acetic acid [10 times the original sample quantity (v/w)].   Thereafter, the sample was centrifuged (4°C, 10,000 × g, 20 min) to collect the supernatants.   Then, acid precipitates were obtained by adding 2 M NaCl to the sample, stirring for 24 h and performing a centrifugation step (4°C, 10,000 × g, 20 min).   The precipitates were then placed in a dialysis membrane (Cell Separation, USA) in cold distilled water to obtain ASC.

Amino acids constituting ASC

ASC extracted from each fish skin was dehydrated with a vacuum freeze dryer (SFDSF-24; Samwon, Korea) and was subjected to the same process described in Section 2.5 for amino acid composition analysis.

Calculation of the ASC ratios of fish skins

To examine the ASC ratios for fish skins from the 4 species, 1 mL dialyzed ASC was kept in a 1.5 mL tube for freeze-drying and subsequent weight comparison. The weights before and after the freeze-drying and the ratios of the dialyzed ASC were then measured. The ratios of ASC extracted from fish skin were calculated as follows:

$S = (Y_1 \times Y_2) / X \times 100$

where, S = ratio of ASC within fish skin (%), Y_1 = weight of acid precipitates before dialysis (g), Y_2 = ratio of ASC to dialyzed and freezedried gel (%) and X = initial wet weight of fish skin (g).

Electrophoretic analysis of RS-AL, ASC, and dried fish meal isolated from fish skins

SDS-PAGE was conducted according to the Laemmli (1970) method in order to determine the molecular weights of fish skin proteins, RS-AL, ASC, and dried fish skin meal. Samples were resuspended at 1 mg/mL in sample buffer (50 mM Tris-HCl, pH 7.5; 50% glycerin, 1% SDS, 0.02% bromophenol blue, BPB) and vortexed. Suspensions were then heated for 5 min at 95°C for heat-induced denaturation, cooled at room temperature for 10 min. The samples were then electrophoresed using a 7.5% gel by using 40% polyacrylamide composed of a 3% stacking gel and a 7.5% separating gel. A mini-protein tetra cell (Bio-Rad Laboratories, Hercules, USA) was used for electrophoresis (200 V, 35 mA/gel). Protein band staining was performed following the method suggested by Fairbanks et al (1971). The SDS-PAGE High Range Molecular Weight Standards (Bio-Rad Laboratories, Hercules, CA) were employed as a marker to identify the molecular weights of samples.

Statistical analysis

Data were analyzed using one-way analysis of variance (ANOVA). When differences were found among the dietary treatments, Duncan's multiple range test was used to compare the mean difference using SPSS software (SPSS, Chicago, IL, USA). Differences were considered significant at $P < 0.05$.

Results

Biochemical characteristics of fish skin

Basic components

Table 1 outlines the results of our analyses of the skins from olive flounder, black rockfish, sea bass, and red sea bream.   Skin moisture levels ranged from 76.11% to 81.10%.   Based on dry matter content, the crude protein levels were highest in red sea bream skin (94.11% ± 0.31%), followed by olive flounder (88.91% ± 3.64%), black rockfish (80.59% ± 1.68%), and sea bass (72.81% ± 0.43%).   However, sea bass had the highest level of crude lipids (25.47% ± 2.59%), whereas red sea bream had the lowest (4.10% ± 1.36%).   The crude lipid levels of black rockfish and olive flounder were 4.10% ± 1.36% and 17.07% ± 0.45%, respectively. The ash levels of black rockfish skin was 10.90% ± 1.87% which is more than 10-fold the levels found in the other 3 species (range - 0.70% to 1.97%).

Table 1. Proximate analysis of skins from olive flounder (*Paralichthys olivaceus*), black rockfish (*Sebastes schlegeli*), sea bass (*Lateolabrax maculatus*), and red sea bream (*Pagurs major*) (%)

	Moisture	*Crude protein	*Crude lipid	*Ash
P. olivaceus	80.87±1.07	88.91±3.64	17.07±0.45	1.97±0.15
S. schlegeli	79.37±0.56	80.59±1.68	5.10±0.71	10.90±1.87
L. maculatus	81.10±2.05	72.81±0.43	25.47±2.59	1.80±0.17
P. major	76.11±1.46	94.11±0.31	4.10±1.36	0.70±0.00

Data are mean ± SD.
*Dry matter.

Fatty acyls

The results of fatty acyl composition analysis are shown in Table 2.   The highest level of saturated fatty acyls was found in red sea bream skin, followed by black rockfish, olive flounder, and sea bass skins.   The composition of saturated fatty acyls was generally C14:0, C16:0, and C18:0; a notable exception was olive flounder skin, which also contained C28:0. Similar levels of skin monoene fatty acyls were found in olive flounder and black rockfish (47.98% and 47.59%, respectively); sea bass and red sea bream both contained more than 40% monoene fatty acyls (41.15% and 44.40%, respectively).   The monoene fatty acyl composition was C16:1, C18:1n9c, C20:1, and C22:1n9; an exception was the absence of C20:1 in sea bass skin.   Polyenes constituted 21.60% of the total in sea bass skin, a proportion that was 2- to 10-fold higher than that of olive flounder, black rockfish, and red sea bream (8.93%, 7.66%, and 2.95%, respectively).   The C18:2n6c configuration of polyene fatty acyl was found in all 4 species.   C20:5 (EPA) and C22:6 (DHA) were also found in olive flounder, and C18:3n3 and C22:6 (DHA) were found in sea bass skin.

Table 2. Fatty acid of skins from olive flounder (*P. olivaceus*), black rock fish (*S. schlegeli*), sea bass (*L. maculatus*), and red sea bream (*P. major*) (%)

Fatty acid	*P. olivaceus*	*S. schlegeli*	*L. maculatus*	*P. major*
C14:0	7.90	6.90	5.29	7.50
C16:0	28.99	29.88	26.15	33.85
C18:0	5.18	7.99	5.82	11.29
C23:0	1.03	-	-	-
Saturated	**43.10**	**44.77**	**37.26**	**52.64**
C16:1	12.60	9.09	12.40	8.57
C18:1n9c	31.09	32.14	26.76	29.03
C20:1	2.26	3.52	1.99	3.65
C22:1n9	2.03	2.84	-	3.15
Monoenes	**47.98**	**47.59**	**41.15**	**44.40**
C18:2n6c	2.23	7.66	12.32	2.95
C18:3n3	-	-	4.17	-
C20:5(EPA)	3.11	-	-	-
C22:6(DHA)	3.58	-	5.11	-
Polyenes	**8.93**	**7.66**	**21.60**	**2.95**
Total	**100.0**	**100.0**	**100.0**	**100.0**
n-3	**6.69**	**-**	**9.28**	**-**
n-6	**2.23**	**7.66**	**12.32**	**2.95**
n-3/n-6	**3.00**	**-**	**0.75**	**-**

Table 3. Amino acid composition of skins from olive flounder (*P. olivaceus*), black rockfish (*S. schlegeli*), sea bass (*L. maculatus*), and red sea bream (*P. Major*) (g/100g)

Amino acid	*P. olivaceus*	*S. schlegeli*	*L. maculatus*	*P. major*
Aspatic acid	3.19	3.36	3.33	3.09
*Threonine	1.39	1.36	1.58	1.42
Serine	2.20	2.51	2.07	2.27
Glutamic acid	5.22	5.22	5.75	4.25
Proline	5.74	5.27	6.94	5.74
Glycine	10.85	10.38	11.38	11.08
Alanine	4.63	4.21	5.16	4.84
Cystine	0.15	0.19	0.11	0.19
*Methionine	1.33	1.45	1.37	1.21
*Valine	1.14	0.99	1.11	1.13
*Isoleucine	0.76	0.81	0.83	0.59
*Leucine	1.64	1.68	1.82	1.42
Tyrosine	0.45	0.46	0.48	0.44
*Phenylalanine	1.34	1.35	1.48	1.30
*Histidine	1.01	1.10	1.09	9.25
*Lysine	1.94	1.95	2.05	1.97
Ammonia	1.61	1.88	2.19	2.03
*Arginine	4.20	4.05	4.59	4.20
Total	**48.79**	**48.22**	**53.33**	**56.42**
*EAA	**14.75**	**14.73**	**15.92**	**22.48**

* Essential amino acid of fin fish: Threonine, Methionine, Valine, Arginine, Histidine, Isoleucine, Leucine, Lysine, Phenylalanine.

Amino acid composition

The amino acid composition of the fish skins shown in Table 3.   Red sea bream skin had the highest total amount of amino acids (56.42 g/100 g), followed by sea bass (53.33 g/100 g), olive flounder (48.79 g/100 g), and black rockfish (48.22 g/100 g).   The highest essential amino acid (EAA) content was found in red sea bream skin (22.48 g/100 g), followed by olive flounder, sea bass, and black rockfish skins (15.92, 14.75, and 14.73 g/100 g, respectively). The content of glycine, a major amino acid of fish skin collagen was highest in sea bass (11.38 g/100 g).   No significant interspecies differences were found, as the contents in red sea bream, olive flounder, and black rockfish were 11.08, 10.85, and 10.38 g/100 g, respectively.

Histological characteristics of fish skin and thickness of fish skin parts

Histological characteristics of fish skin

The histological characteristics of dorsal and ventral skin parts from each of the 4 species are shown in Figures 1 and 2.   The general structure of the dorsal skin part of olive flounder was composed of epidermal and dermal layers (Fig. 1-A).   Mucous cells containing acidic mucopolysaccharides were found in secretory cells of the epidermal layer.   The dermal layer was composed of collagen fiber, whereas the muscle layer contained muscular fibers (Figs. 1-A-1 and 2).

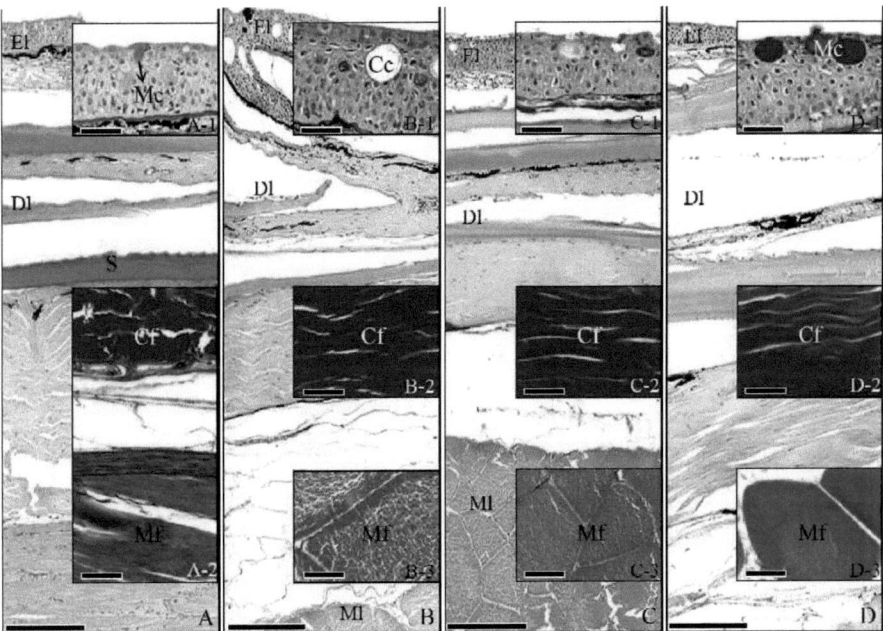

Fig. 1.   Microscopic observation of dorsal skin parts in 4 fish species. A: olive flounder (*P. olivaceus*), H&E stain, A-1; AB-PAS (pH 2.5) reaction, A-2; Masson's trichrome stain. B: black rockfish (*S. schlegeli*), H&E stain, B-1; AB-PAS (pH 2.5) reaction, B-2 and 3; Masson's trichrome stain. C: sea bass (*L. maculatus*), H&E stain, C-1; AB-PAS (pH 2.5) reaction, C-2 and 3; Masson's trichrome stain. D: red sea bream (*P. major*), H&E stain, D-1; AB-PAS (pH 2.5) reaction, D-2 and 3; Masson's trichrome stain. Cc: club cell, Cf: collagen fiber, Dl: dermal layer, El: epidermal layer, Mc: mucous cell, Mf: muscle fiber, Ml: muscle layer, S: scale. Scale bars in the figure and the inset indicate 50 μm and 25 μm, respectively.

Similarly, the ventral skin parts of olive flounder were also composed of dermal and epidermal layers (Fig. 2-A). Mucous cells (containing neutral mucopolysaccharides) and pigment cells were the predominant type of secretory cells found in the epidermal layer (Fig. 2-A-1). The dermal layer was also composed of collagen fibers (Fig. 2-A-2), and the muscle layer comprised well-developed muscle fibers (Fig. 2-A-3).

The dorsal and ventral parts of black rockfish were composed of epidermal and dermal layers (Figs. 1-B and 2-B). The dermal layer of the dorsal parts contained collagen fibers and some muscular fibers, and the dorsal parts had well-developed collagen fibers (Figs. 1-B-1, 2, 3 and 2-B-1, 2, 3).

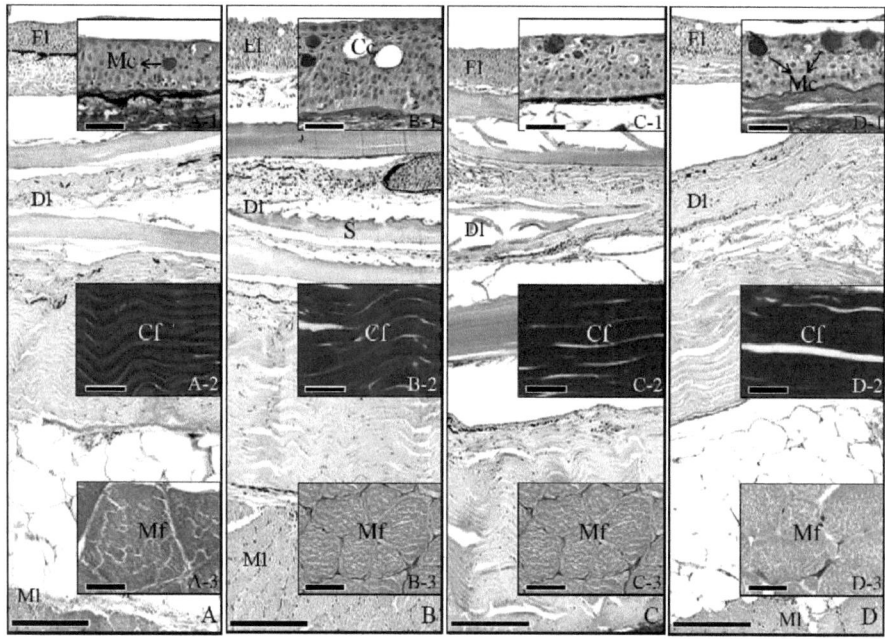

Fig. 2. Microscopic observation of ventral skin parts in 4 fish species. A: olive flounder (*P. olivaceus*), H&E stain, A-1; AB-PAS (pH 2.5) reaction, A-2 and 3; Masson's trichrome stain. B: black rockfish (*S. schlegeli*), H&E stain, B-1; AB-PAS (pH 2.5) reaction, B-2 and 3; Masson's trichrome stain. C: sea bass (*L. maculatus*), H&E stain, C-1; AB-PAS (pH 2.5) reaction, C-2 and 3; Masson's trichrome stain. D: red sea bream (*P. major*), H&E stain, D-1; AB-PAS (pH 2.5) reaction, D-2 and 3; Masson's trichrome stain. Cc: club cell, Cf: collagen fiber, Dl: dermal layer, El: epidermal layer, Mc: mucous cell, Mf: muscle fiber, Ml: muscle layer, S: scale. Scale bars in the figure and the inset indicate 50 μm and 25 μm, respectively.

The skin of sea bass was composed of epidermal, dermal, hypodermal, and muscle layers. The types of secretory cells found in the epidermal layer included mucous cells and club cells (Figs. 1-C-1, 2, 3 and 2-C-1, 2, 3). The dermis layer was composed of well-developed collagen cells in both dorsal and ventral parts, and the muscle layer was composed of muscle fibers (Figs. 1-C and 2-C).

The dorsal and ventral skin parts of red sea bream were composed of epidermal, dermal, hypodermal, and muscle layers. There was a paucity of secretory cells in the epidermal layer (Figs. 1-D-1, 2, 3 and 2-D-1, 2, 3). Muscle layers were composed of well-developed muscle fibers in both dorsal and ventral parts (Figs. 1-D and 2-D).

Thickness of fish skin parts

We measured each layer of the dorsal and ventral skin parts. The dorsal cutis of the olive flounder was 540.24 μm thick (epidermal layer, 49.23 μm; dermal layer, 491.01 μm), while the ventral cutis was considerably thicker (736.92 μm; epidermal layer, 43.08 μm; dermal layer 693.84 μm). A similar pattern was also seen for black rockfish; dorsal and ventral skins were 367.10 μm thick (epidermis, 20.13 μm; dermis, 346.97 μm) and 528.48 μm thick (epidermis 23.42 μm; dermis 505.06 μm), respectively. In red sea bream, a substantial difference in thickness was observed between the dorsal (287.45 μm) and ventral skins (407.60 μm). This was most pronounced in the epidermal layer, as the ventral part was 3-fold thicker than the dorsal section (67.60 μm and 18.52 μm, respectively). The dermal layer was also thicker in the ventral part (340.00 μm) than in the dorsal part (268.93 μm). The values for sea bass were similar to the other 3 species, since the thickness of the dorsal and ventral skins were 485.44-μm (epidermis, 72.38 μm; dermis, 413.06 μm) and 490.28-μm (epidermis 75.82 μm; dermis, 414.46 μm), respectively. Olive flounder had the thickest ventral cutis, and sea bass had the thickest dorsal and ventral epidermis layers. Olive flounder also had the thickest dermis, which contained a high level of collagen, whereas the dermis of red sea bream was the thinnest.

ASC ratios and molecular biochemical characteristics of fish skin.

ASC ratios within fish skin

Based on wet weight, the ASC ratio of red sea bream skin (9.78%) was highest, followed by olive flounder (7.82%), sea bass (5.64%), and black rockfish (2.53%). Based on dry weights, the ASC ratio of the olive flounder skin (20.69%) was highest, followed by that of the red sea bream (20.44%), sea bass (14.74%), and black rockfish (11.00%).

ASC amino acid composition

The ASC amino acid composition of skins from olive flounder, black rockfish, sea bass and red sea bream were analyzed. The values for olive flounder, black rockfish, sea bass and red sea bream were 69.32, 72.04, 74.77, and 66.25 g/100 g, respectively, confirming that sea bass had the highest abundance of ASC. The concentration of glycine, an identifier of the repetitive Gly-X-Y amino acid sequence within the triple helical domain of collagen, was 17.47, 17.88, 18.56, and 16.86 g/100 g for olive flounder, black rockfish, sea bass, and red sea bream, respectively. This accounted for approximately one-third of the total amino acids for each species. The skin of each species had a high proline content: 10.75 g/100 g in olive flounder, 10.74 g/100 g in black rockfish, 12.59 g/100 g in sea bass, and 10.29 g/100 g in red sea bream.

SDS-PAGE of fish skin, RS-AL, ASC, and dried fish skin meal

Figure 3 shows the results of our electrophoretic analysis of fish skins and the associated derivatives.

For all 4 species, a subunit structure of α 1(I), α2(I) together with a dimer β-chain was found in skins, RS-AL, ASC, and fish meal. In all cases, a typical I type SDS-PAGE pattern with 2 high molecular bands on top of the β-chain was found. Molecular weight analysis of the skins, RS-AL, ASC, and fish meal revealed the presence of α1(I) and α2(I) protein bands at ~116.2 kDa, and a major β-chain protein band at ~200 kDa. Notably, the protein structures of skins, RS-AL, and ASC were maintained after heat-drying.

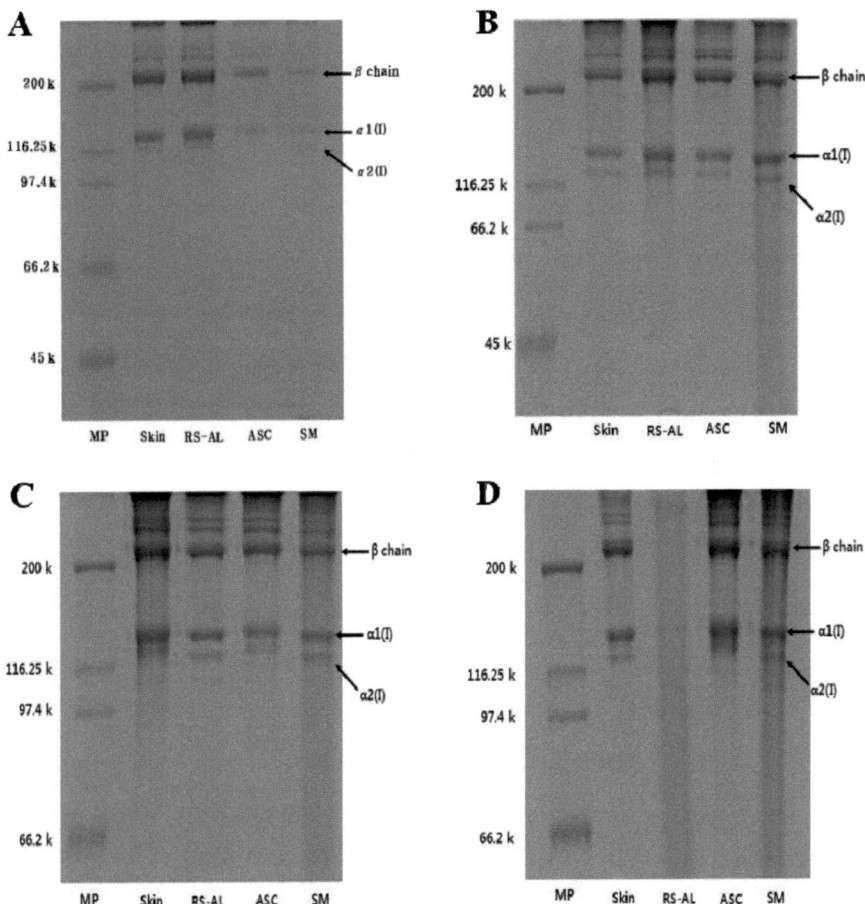

Fig. 3. SDS-PAGE patterns of marker protein (MP), skin, residues after alkali extraction (RS-AL), acid-soluble collagen (ASC), and skin meal (SM) of olive flounder (A), black rockfish (B), sea bass (C), and red sea bream (D).

Discussion

In the preparation of sashimi or sushi from marine fishes such as olive flounder, black rockfish, sea bass, and red sea bream, only the muscle tissue is used. Thus, virtually all the other fish tissues, including the head parts, bones, guts, and skins are left as by-products. The fish skin in particular has a very high protein composition (90%), with a variety of high molecular weight proteins such as collagen, albumin, myosin, and elastin present at high concentrations (Arvanitoyannis & Kassaveti, 2008; Bechtel, 2003; Bechtel & Smiley, 2010; Jayathilakan et al., 2012). Therefore, we designed this study to determine whether fish skin should be considered as a potentially protein-rich alternative food source.

Olive flounder, black rockfish, sea bass, and red sea bream had a high crude protein content (ranging from 72–97%). This level is similar to that found in animal feeds including casein, gelatin, and blood meal, which contain up to 80% protein. Fish skin crude protein content is higher than that of fish meal (60–70% crude protein content; NRC, 1993), a major protein source for assorted feed for sea fish (Rathbone et al., 2001). Thus, we suggest that fish skin could be a replacement protein source for the more expensive fish meal that is currently used.

Notably, ash content is less than 2% in olive flounder, sea bass, and red sea bream. Phosphorous, one of the main components of ash, is excreted from the body owing as it is not efficiently metabolized, and this causes eutrophication of water (Watanabe, 1991). In this context, feeds prepared from fish skin for farmed fish may reduce environmental contamination.

Fish cannot synthesize linoleic acid (C18:2n-6) and linolenic acid (C18:3n-3) like other vertebrates; therefore, essential fatty acids (EFAs) must be supplied in fish food (NRC, 1993). Moreover, fish have an innate ability to convert 18-carbon highly unsaturated fatty acids (HUFAs) into longer and higher HUFAs (Owen et al., 1975). Thus, fish EFA composition is dependent on the precise conversion mechanisms that occur through fatty acid metabolism (NRC, 1993). The EFA requirement varies depending on the environments of fish habitats. For example, freshwater fish need either linoleic (C18:2n-6), linolenic acid (18:3n-3), or both, whereas stenohaline fish require HUFAs, such as eicosapentanoic acid (C20:5n-3, EPA) and docosahexaenoic acid (C22:6n-3, DHA), directly in the diet (NRC, 1993). Here, we show that the 4 fish species under investigation contained very high monoene (but low polyene)

17

concentrations.   In particular, n-3 fatty acids, including essential fatty acids such as EPA and DHA, were virtually absent, but C18:2n6c, which is indispensable for freshwater fish, was found in the skins of all 4 fish.   Fish meal, which is the main farmed fish feed, already contains a large amount of essential fatty acids such as HUFAs.   However, fish oil supplements (such as squid liver oil and cod-liver oil that have high n-3 HUFA content) are added in order to increase the energy level and supply a sufficient amount of essential fatty acids (Barrows & Hardy, 2000).   Therefore, the fatty acid deficit in the skins of olive flounder, black rockfish, sea bass, and red sea bream can easily be complemented by adding lipid sources rich in HUFA.

The balance of amino acids in farmed fish feeds is important, as it directly affects the nutritional quality of constituent proteins (NRC, 1993).   Essential amino acids (EAAs) for fish include arginine, histidine, isoleucine, leucine, lysine, methionine, phenylalanine, threonine, tryptophan, and valine.   Lack of EAA supply can lead to growth deficiency in fish (Wilson, 2002).   The balanced EAA composition of fish meal is critical for marine fish, and alternative protein sources generally have insufficient amounts of EAA (Lee & Lee, 1996). The EAA composition of the skin from olive flounder, black rockfish, sea bass, and red sea bream ranged from 14.73–22.48 g/100 g.   This is 50–70% of the reported EAA content for white fish meal (31.92 g/100 g) and brown fish meal (30.25 g/100 g).   Thus, fish skin cannot completely substitute fish meal as a feed for farmed fish as it does not offer the required amount of EAAs.   We suggest that a species-specific amount of amino acids could be added to the fish skin extract, or the extract could be mixed with fish meal in order to use fish skin as protein source.   The proline and glycine composition of fish skin was 2- to 3-fold that of white fish meal.   Proline and glycine are classified as nonessential amino acids, but they are important for maintaining collagen structure (Gly-X-Y).   Moreover, Kim et al (2002) reported that adding proline and glycine to feed prepared from black rockfish fry boosts growth and feeding efficiency.   Thus, we infer that proline and glycine content is an important determinant of suitability of fish skin when used as as fish feed.

18

Generally, fish skin tissues contain special sensory organs such as the lateral line system and taste buds, which play a crucial role as the outermost defense wall against the environment (Her et al., 2002).   Although the precise composition of fish skin varies between species, is basically divided into 2 layers: the epidermis and dermis.   The basement membrane divides the epidermal and dermal layers, and the subcutaneous layer lies between the dermal and muscle layers (Muyonga et al., 2004).   In particular, the dermis layer has a high concentration of collagen fibers.   We confirmed the presence od collagen fibers in the dorsal and ventral dermis layers of all 4 fish species studied herein.   In particular, olive flounder had thicker dorsal and ventral parts than the other species, and its dermal layer accounted for more than 90% of the skin; this indicates that it is the optimal protein source among the 4 species examined.

Red sea bream had the highest ASC ratio based on wet matter, but olive flounder had the highest ratio in terms of dry matter.   The high ASC extraction ratio in olive flounder is likely because it has the thickest skin of the 4 species.   A greater amount of ASC was extracted from red sea bream in comparison with black rockfish and sea bass, presumably due to differences in the size and thickness of the scales.   Sea bass contained the highest levels of amino acids, indicating that the total amount of amino acids is not directly associated with the thickness of fish skin.

Kim, Kim, and Cho (1997) performed SDS-PAGE analysis of the molecular weights of soluble collagen in the skins of yellowfin sole (*Limanda aspera)*, red cod (*Physiculus bacchus*), Pacific cod (*Gadus macrocephalus*), Alaska pollock (*Theragra chalcogramma*), and red squid (*Ommastrephes bartrami*).   They found that soluble collagen is composed of an α-chain and a dimer; that is, a β-chain is present with an α-chain of at least 116 kDa molecular weight.   In the case of red cod and red squid, the α1-chain and α2-chains reportedly molecular weight slightly higher than 116 kDa, and the β-chain has a molecular weight around 205 kDa.   Here we found that all 4 species contain collagen that is composed of an α-chain and its partner β-chain.   α1-chain and α2-chains migrating above 100 kDa were found, whereas the β-chain migrated around 200 kDa.   Notably, fish skin, RS-AL, and ASC have SDS-PAGE patterns of typical I type collagen without denaturation, as compared with fish meal after heat-drying. This finding suggests that hot-air-dried fish skin is an ideal food source owing to its content of proteins and, specifically, of collagen.

Animal by-products are unsuitable as an alternate food source to fish meal, as they have imbalances in EAAs, low protein digestibility, and are subject to spoilage (Dong et al., 1993). Furthermore, the feed source must contain more than 35% protein, have low anti-protease activity, have an amino acid composition similar to that of fish meal, and yield less than 15% ash, and 0.5% or more soluble phosphorous (Chalamaiah et al., 2012).   The fish examined in this study have high crude protein contents.   The abundance of collagen fibers and ASC content found in the dermis of fish skin, as well as the resistance to denaturation exhibited by collagen in hot-air-dried fish skin, suggests that fish skin may be used as a protein source. However, while this indicates that their skin is suitable as a protein source for farmed fish, the low content of essential fatty acids and EAAs means that fish skin cannot be used as a 'standalone' replacement for current fish meals.   We suggest that the optimal formulation for new fish meals will feature a combination of traditional food sources and components of fish skin.

Chapter 2

Dietary flounder skin improves growth performance, body composition, and stress recovery in the juvenile black rockfish (*Sebastes schlegeli*)

This study investigated the relationship between flounder skin meal (FSM) and vitamin C in mediating collagen biosynthesis. Based on the vitamin C requirements (150 mg/kg) of the black rockfish (mean body weight 10.05 ± 0.44 g), a vitamin C level of 400 mg/kg was selected, and 0, 5, 10, or 20% of the casein (purified proteins) in the diet was replaced with FSM. The feeding study was conducted for 8 weeks by using 4 experimental groups. The FSM supplementation resulted in improvement in growth performance, decrease of body lipids. Furthermore, it elevated the HDL-cholesterol levels and total protein content, reduced blood lipids, and led to rapid recovery in stress, which confirm the functionality of FSM with high collagen content.

Introduction

Numerous studies have been performed investigated vegetable and animal proteins that could replace fishmeal in fish feed. In particular, there have been many studies on the use of vegetable protein sources such as soybean meal (Lee & Jeon, 1996; Murai et al., 1982b; Robert et al., 1993), cottonseed meal, and rapeseed meal (Lee & Yoo, 1996; Pham et al., 2005), which have a relatively stable supply compared to fishmeal, to replace fishmeal as a source of protein. However, plant resources are constantly in competition with livestock and human consumption, and the recent development of plant extract fuels such as bioethanol will eventually lead to an increase in the price of the plant resources usable as protein resources (Rena &

Hasan, 2009). The by-products of the processing of terrestrial livestock such as cows, chickens, and pigs could be used as animal protein sources, since they have a relatively high protein content and qualitatively similar amino acid composition to fishmeal, and are inexpensive and stably supplied. Various studies have been conducted on their use as protein sources to replace fishmeal in fish feed (Ai et al., 2006; Bai et al., 1998; Kikuchi et al., 1994; Kim & Bai, 1997, 1999; Lee & Lee, 1998; Sato & Kikuchi, 1997). However, the rise of safety issues due to serious infectious diseases like mad cow disease, swine fever, and avian influenza has gradually restricted the use of livestock by-products lately. Thus, as there are economic and safety issues with using terrestrial protein sources to replace fishmeal, securing economic and safe protein sources from marine products rather than terrestrial products is necessary. Many researchers have investigated by-products obtained from processing marine animal as potential protein sources, including shrimp by-products (Cruz-Suárez et al., 1993), tuna muscle by-products (Uyan et al., 2006), shrimp and fish by-products (Li et al., 2004), squid liver meal mixing soybean meal with by-products of squid processing (Kim & Bai, 1997), fish bone and crab by-products (Goytortúa-Bores et al., 2006; Lee et al., 2010; Toppe et al., 2006), and fish by-products (Foster et al., 2005). Of the fishery by-products, even though fish skins obtained from the consumption of raw fish are a good protein source high in collagen content, by-products such as bones and internal organs are only partially used and mostly discarded.

Collagen, a main component of the extracellular matrix, is distributed in multicellular animals and is a protein that builds skeletons for the body structure. It also has many functions such as morphological changes and biological defenses in the proliferation, differentiation, and development of the cells involved in the maintenance of body or organ structures (Eyre & Wu, 1987; Fessler & Fessler, 1987; Kuhn, 1987). In addition to type I collagen found in the dermis, bones, fins, scales, and muscles; type II found in notochords; type XI found in cartilage; and type V found in muscles, scales, and skin, a number of molecular varieties have been found in fish such as truncus (C-B) collagen present in the muscles and intestinal canals of cyclostomes, a special type of fish (Kimura, 1997; Yoshinaka, 1989). Until recently, collagen that originated from terrestrial animals and gelatin, the denatured products thereof, were considered to be safe and useful substances in various industry fields such as food, clothing, cosmetics, and cell culture. However, due to safety concerns following frequent outbreaks of mad cow disease, swine fever, and avian influenza, "marine collagen" derived from marine animals has gained attentions as a substitute (Lupi, 2002; Schrieber & Seybold, 1993). Recently, many researchers have investigated the physico-chemical properties of collagen and gelatin extracted from fish skin by-products of *Astroconger myriaster*, *Navodon modestus*, *Loligo bleekeri*, *Todarodes pacificus*, *Limanda aspera*, *Gadus macrocephalus tilesius*, *Thunnus albacares*, *Physiculus bacchus*, *Theragra chalcogramma*, and *Isurus oxyrinchus* to evaluate their industrial feasibility (Kim, 1996; Kim & Cho, 1996; Kim & Kwak, 1991; Kim et al., 1993a, 1993b, 1996, 1997; Kwon et al., 2008; Park et al., 2009; Yoo et al., 2008).

In marine animals, the provision of feed mixed with hydroxyproline, a main amino acid of collagen, with vegetable proteins improved the growth of *Salmo salar* L. and increased the hydroxyproline content in the vertebral column (Aksnes et al., 2008). Consequently, positive *in vivo* functional effects of collagen depend upon the ingestion of its hydrolysates, peptides, or hydroxyproline and using fish skin containing collagen in abundance can provide these functional effects. Collagenase activity, especially in teleosts, was observed to be high in pancreatic tissues (Yoshinaka et al., 1978), indicating that the direct digestion of collagen is possible and collagen intake in fish is expected to have various physiological effects like those described above for terrestrial vertebrates.

Vitamin C, an essential cofactor in the production of hydroxyproline and hydroxylysine, which are important for collagen generation (Sandel & Daniel, 1988), is known to be an essential nutrient with various functions in fish (Bai et al., 1996). Vitamin C also governs the formation of pyridinoline, which is known to cross-link materials during collagen maturation (Chan et al., 1990; Kim, 2004; Nurad et al., 1981). In particular, the addition of vitamin C is reported to increase intracellular collagen synthesis when culturing 3 T6 fibroblast cells in mice (Kim, 2006), indicating that vitamin C is significantly related to collagen. In fish skin, especially where there is a high content of the amino acids important for collagen composition like proline and glycine, adding vitamin C is anticipated to provide interactive outcomes with the collagen components in the fish skin.

Therefore, this study evaluated the usefulness of fish skin by-products as protein sources in mixed feed for fish farming. Based upon the results of previous studies, among *Paralichthys olivaceus*, *Sebastes schlegeli*, *Lateolabrax maculatus*, and *Pagrus major* (Cho et al., 2014), the skin of *Paralichthys olivaceus* was the thickest and was high in collagen fiber content and ASC. Therefore, it was powdered after hot air drying without special processes and directly provided to the black rockfish, and its potential as a protein source was investigated by elucidating the physiological characteristics of the fish. We also confirmed the synergistic effects of the collagen ingestion from the fish skin and vitamin C addition.

Material and Methods

Experimental diet preparation

The skin of *Paralichthys olivaceus*, which has the highest farming yield and raw fish consumption in Korea, is easy to secure in large quantities due to its low use, thickness, and high content of collagen fibers and ASC, and was obtained from nearby fish markets. The fish skin was washed with fresh water and was subjected to hot air drying (50–60°C) followed by grinding via a high speed grinder (ZM-1000, Retsch Co., Japan) to prepare flounder skin meal (FSM).

Ingredients and proximate compositions of the experimental diets in response to FSM supplementation and the results of vitamin C analyses are shown in Table 4. Proximate analyses were carried out to evaluate the nutritional composition of the prepared diets, and the vitamin C content of the diets were analyzed using the 2,4-dinitrophenyl hydrazine (DNP) colorimetric method. Briefly, 5% metaphosphoric acid solution was added to a certain amount of the sample and ground by a homogenizer. The sample was then centrifuged and 5% metaphosphoric acid solution was added to the supernatant to make a final volume of 100 mL. One to 2 drops of 0.2% indophenol solution were added to 2 mL of the prepared solution and then 2 mL of thiourea-metaphosphoric acid solution and 1 mL of DNP solution were added sequentially followed by reaction at 50°C for 30 min. The sample was cooled down and 5 mL of 85% sulfuric acid solution was added. Absorbance was then measured at 540 nm and vitamin C was quantified by a standard calibration curve.

FSM is a high protein meal containing more than 80% crude proteins but is lacking in essential amino acids compared to fishmeal. When fishmeal is replaced by FSM, unknown factors present in essential amino acids and fishmeal may affect the experimental fish, making it difficult to evaluate the influences of FSM supplementation on experimental fish. Therefore, to maintain proper balances of the essential amino acids and to minimize the effects of the unknown factors in the fishmeal, white fishmeal (FF Skagen LT Supreme, Denmark) was fixed at the same level throughout the experimental diets. Casein, a purified protein, was used to control the protein content of each experimental diet. Squid liver oil (Ihwa, Korea) rich in DHA and EPA, which are essential fatty acids for the black rockfish, was used as a lipid source. Flours (CJ, Korea) and α-starch were employed as carbohydrate sources to control energy and bind the diets. To find the relationship between the level of vitamin C (Sandel & Daniel, 1988), and FSM supplementation on the fish body, vitamin C (400 mg/kg) was added based upon the vitamin C requirements of the black rockfish reported by Bai et al (1996). There were four experimental groups, including a control group with fishmeal and casein only and three experimental groups with 5, 10, or 20% of the casein replaced by FSM. The experimental diet was prepared 5 mm in diameter with a moist pellet maker (Sun Brand Industrial, Korea) and stored at −45°C until further utilization.

Feeding trial, growth performance, and proximate analysis

Experimental fish were produced from a private pond hatchery in March 2010 and transferred to the Fisheries Science Institute, Chonnam National University, Korea. Black rockfish (mean body weight 10.05 ± 0.44 g, mean body length 8.28 ± 6.94 cm) fed commercial diets (Kurosoi, CP48%, CL12%, Japan) for 3 months were used in the study. The fish were transferred to a 300 L square FRP water tanks, each containing 40 fish, with treatments in triplicate. The diet was given twice a day (08:00, 18:00) ad libitum for 8 weeks. Water that was filtered through a high-pressure sand filter was provided in a running water system at 5 L/min. During the feeding, the fish were kept at a water temperature of $20.2 \pm 2.3°C$, dissolved oxygen of 6.3 ± 0.4 mg/L, and salt concentration of $32.0 \pm 1.2‰$. The procedure was reviewed and approved by the Chonnam National University Institutional Animal Care and Use Committee (CNU IACUC-2010-45).

Table 4. Ingredients and proximate composition of experimental diets with various levels of FSM

Ingredient (%, DM)	FMS level (%)			
	Control (0)	5	10	20
White fish meal	41	41	41	41
Casein	20	15	10	0
Flounder skin meal (FSM)	0	5	10	20
L-ascorbic acid	0.04	0.04	0.04	0.04
Wheat flour	22.56	23.26	23.96	25.26
Squid liver oil	8.4	7.7	7	5.7
α-potato starch	3	3	3	2
Vitamin premix[a] (vitamin C free)	2	2	2	2
Mineral Premix[b]	2	2	2	1
Choline Chloride	1	1	1	1
Total	100	100	100	100
Vitamin C in diets	391.35	411.37	416.77	409.56
Proximate analysis				
Protein	45.71	45.91	47.51	46.51
Lipid	10.11	9.21	10.91	9.91
Ash	8.01	8.21	8.31	8.31

Data are mean ± standard deviation (SD) of three group of fish (n = 3). Values with different superscripts are significantly different ($P < 0.05$)

DM dry matter

ns not significant

[a] Vitamin premix (mg/g mixture) : retinol acetate, 0.81mg; cholecalciferol, 0.012mg; vitamin E, 22.5mg; vitamin K_3, 2.5mg, thiamine, 5.5mg; riboflavin, 10mg; pyridoxine, 6mg; niacin, 37.5mg; folic acid, 2mg; biotin 0.05mg; inositol 50mg. All ingredients were diluted with alpha-cellulose to 1g.

[b] Mineral premix (mg/g mixture) : Mn, 3.2mg; Zn, 3.2mg; Fe, 3.0mg; Cu, 0.36mg; $MgSO_4$, 100mg; KCl (47%), 60mg; $Al(OH)_3$, 1.06mg; $Ca(IO_3)_2$, 0.475mg; $CoSO_4$, 0.475mg. All ingredients were diluted with alpha-cellulose to 1g.

*Dry matter

Fish bodies were measured before and after the experiments. The experimental fish were starved for 24 h prior to the measurement and the whole body weight of the fish was determined under anesthesia with 100 ppm of the anesthetic for fish only (AQUI-S, New Zealand). In addition, 10 fish randomly picked from each water tank at the beginning and end of each experiment were subjected to body weight, whole length, and body length measurement to calculate the weight gain (WG), feed efficiency (FE), and protein efficiency rate (PER), as growth performance indicators. To investigate the effects of the diet with FSM supplementation on fish body internal organs, the hepatic weight, the viscera weight, and the gut length were determined after dissection, and the viscerasomatic index (VSI), hepatosomatic index (HSI), stomach somatic index (SSI), relative length of the gut (RLG), and condition factor (CF) were calculated via the equations below. The livers and dorsal muscles were isolated for component analyses and then stored at $-45°C$ until further analyses.

VSI (%) = viscera weight (g) $\times$ 100/body weight (g)

HSI (%) = hepatic weight (g) $\times$ 100/body weight (g)

SSI (%) = stomach weight (g) $\times$ 100/body weight (g)

RLG = gut length (cm)/total length (cm)

CF = body weight (g) $\times$ 100/body length (cm^3)

Proximate analyses of the experimental diets, livers, dorsal muscles, and whole body were carried out based on AOAC methods (2002). Moisture content was measured using a moisture analyzer with an air-oven method (HR 73 halogen moisture analyzer, Switzerland), and crude protein content was determined by a Kjeldahl nitrogen quantification method (N $\times$ 6.25) using an automated analyzer (KJELTEK auto sampler system 1035 analyzer, Switzerland). Crude fat content was analyzed by an ether extraction method via an auto extraction unit (Soxtec 2050 auto extraction unit, Switzerland), and crude ash content was measured using a direct furnace method at 550°C (EYELA Electric furnace TMF-3100, Japan).

Hematological analysis

Blood composition was analyzed to investigate the health of the experimental fish. After not eating for 24 h at the end of the experiments, blood was collected from the tail blood vessels of 10 fish that were randomly picked in each group using a 1 mL disposable syringe treated with heparin-Na (Sigma, 100,000 units, 2.5 mg/mL), as an anticoagulant. Hemoglobin (Hb, Asan Pharm, Korea) was then measured immediately in whole blood using a commercial kit, and hematocrit (Ht) was measured using a glass capillary. The collected blood was centrifuged for 15 min (12,000 rpm, 4°C), and the separated plasma was subjected to analyses of glucose, total cholesterol, high density lipoprotein (HDL)-cholesterol, glutamic oxaloacetic transaminase (GOT), glutamic pyruvic transaminase (GPT), total protein, and triglyceride using a kit (Asan Pharm, Korea).

Lysozyme activity was measured using a turbidimetric method based on Parry et al (1965). Briefly, 950 μL of a *Micrococcus lysodeikticus* (0.2 mg/mL) suspension (pH 6.2) was mixed with 50 μL of serum and was reacted at 25°C for 30 s and 4.5 min followed by measuring absorbance at 530 nm. Lysozyme activity was expressed as units/mL, with 1 unit indicating a decrease in absorbance of 0.001/min.

Stress recovery tests

Based on the study by Ji et al (2009), the stress recovery rate was measured by an anesthesia test and an air exposure test within 48 h. For the anesthesia test, 10 fish were randomly picked in each experimental group and were given anesthesia in the water bath with 800 ppm of 2-phenoxyethanol (Sigma, USA) for 3 min. The fish were then transferred to running water and the recovery time was measured every 30 s by using a timer, in triplicates for each experimental group. For the air exposure test, 10 fish in each experimental group were randomly collected in a square plastic basket that strains out water and were exposed to the air for 25 min. The fish were then transferred to running water and mortality was measured after 6 h, in triplicates for each experimental group.

To investigate blood composition changes with respect to air exposure stress, 10 fish in each experimental group were randomly collected in a square plastic basket that strains out water and were exposed to the air for 5 min. After being transferred to running water, 2 fish in each experimental group at 1, 2, 4, and 6 h post-treatment were treated with 100 ppm of an anesthetic for fish (AQUI-S, New Zealand) and blood was collected from the tail blood vessels. The blood was centrifuged for 15 min (12,000 rpm, 4°C) and plasma was separated followed by analyzing hematocrit, hemoglobin, glucose, total cholesterol, triglyceride, GOT, and GPT via a commercial kit (Asan Pharm, Korea).

Statistical analysis

Statistical analyses of the results were performed using an ANOVA-test. The significance of the means was tested by a Duncan's multiple range test (Duncan, 1955) in the SPSS statistical program.

Results

Growth performance

The growth performance of the black rockfish fed diets with different levels of flounder skin meal (FSM) for 8 weeks is shown in Table 5. The survival rate was 100% in all groups. WG and SGR were significantly greater for FSM 20% than the control ($P < 0.05$). The feed intake (FI) was significantly greater for FSM 20% than the control ($P < 0.05$). For FE, no significant differences were found between the control group and the FSM supplement groups, and FSM 20% was observed to be significantly higher than FSM 10% ($P < 0.05$). The protein intake (PI) was significantly greater for FSM 20% than the control ($P < 0.05$). The protein efficiency rate (PER) was not significantly different among any of the experimental groups.

Table 5. Growth performance of black rockfish, *S. schlegeli* fed the test diets with various levels of FSM for 8 weeks

	FSM level (%)			
	Control (0)	5	10	20
IBW (g)[1]	10.90±0.37	10.60±0.12	10.47±0.17	10.92±0.45
FBW (g)[2]	33.12±1.54^{ab}	32.22±1.72^{ab}	30.56±0.64^{a}	33.72±1.98^{b}
SR (%)[3]	100^{ns}	100	100	100
WG (%)[4]	185.35±6.06^{a}	203.87±14.37^{ab}	191.98±10.51^{ab}	208.71±6.97^{b}
SGR[5]	0.36±0.02^{a}	0.39±0.03^{ab}	0.36±0.01^{a}	0.41±0.03^{b}
FI (g)[6]	31.12±0.45^{a}	31.76±1.02^{a}	31.17±0.45^{a}	36.58±1.03^{b}
FE (%)[7]	60.23±3.56^{ab}	64.40±4.88^{ab}	59.85±2.35^{a}	67.94±4.65^{b}
PI (g)[8]	17.43±0.25^{a}	17.78±0.57^{a}	17.45±0.25^{a}	20.31±0.59^{b}
PER[9]	1.16±0.06^{ns}	1.21±0.06	1.15±0.04	1.12±0.05

Data are mean ± standard deviation (SD) of three group of fish (n = 45). Values with different superscripts are significantly different ($P < 0.05$)

ns not significant

[1] Initial mean body weight

[2] Final mean body weight

[3] Survival rate

[4] Weight gain: (final body weight - initial body weight/initial body weight) × 100

[5] Specific Growth rate (%): (final body weight - initial body weight)/the time interval in days × 100

[6] Feed intake

[7] Feed efficiency (%): (fish weight gain/feed intake) × 100

[8] Protein intake

[9] Protein efficiency rate (%): (fish weight gain/protein intake) × 100

The CF, VSI, HSI, SSI, and RLG of the black rockfish fed diets with different levels of FSM for 8 weeks are shown in Table 6. Significant differences were not observed in CF, VSI, HIS, SSI, and RLG between the control group and the FSM supplement groups.

Table 6. Condition factor (CF), viscerasomatic index (VSI), hepatosomatic index (HSI), stomach somatic index (SSI), and relative length of gut (RLG) of black rockfish, *S. schlegeli* fed the test diets with various levels of FSM for 8 weeks

	FSM level (%)			
	Control (0)	5	10	20
CF	3.11 ± 0.08^{ns}	2.95 ± 0.27	2.88 ± 0.20	2.91 ± 0.33
VSI	10.76 ± 1.27^{ns}	11.29 ± 0.90	11.48 ± 3.03	11.01 ± 1.20
HSI	3.35 ± 0.61^{ns}	3.61 ± 0.35	3.56 ± 0.76	3.07 ± 0.22
SSI	1.00 ± 0.19^{ns}	1.12 ± 0.17	1.07 ± 0.07	1.22 ± 0.53
RLG	0.89 ± 0.14^{ns}	0.93 ± 0.08	0.94 ± 0.13	0.97 ± 0.09

Data are mean ± standard deviation (SD) of three group of fish (n = 10). Values with different superscripts are significantly different ($P < 0.05$)

ns not significant

Proximate analysis

Proximate analysis results of livers, dorsal muscles, and whole fish bodies are shown in Table 7.

Whole body moisture showed no significant differences between the control group and the FSM supplement groups, while FSM 20% was significantly higher than the FSM 10% group ($P < 0.05$). Liver moisture in FSM 20% was significantly higher than in the control, FSM 5%, and FSM 10% groups ($P < 0.05$). In the dorsal muscles, there were no significant differences between the control group and the FSM supplement groups.

Table 7. Proximate analysis of whole body, liver, and dorsal muscle of black rockfish, *S. schlegeli* fed the test diets with various levels of FSM for 8 weeks (wet matter)

		FSM level (%)			
		Control (0)	5	10	20
Moisture (%)	whole body	69.07 ± 0.35^{ab}	68.82 ± 0.58^{ab}	68.26 ± 0.68^{a}	69.51 ± 0.63^{b}
	liver	47.55 ± 1.37^{a}	46.00 ± 0.13^{a}	46.26 ± 0.25^{a}	51.72 ± 1.18^{b}
	dorsal muscle	72.79 ± 2.17^{ns}	70.47 ± 0.78	73.03 ± 1.44	71.74 ± 1.08
Crude protein (%)	whole body	14.47 ± 0.42^{b}	14.73 ± 0.05^{bc}	15.08 ± 0.01^{c}	14.06 ± 0.05^{a}
	liver	8.37 ± 0.06^{a}	8.74 ± 0.12^{b}	8.77 ± 0.05^{b}	8.67 ± 0.01^{b}
	dorsal muscle	18.13 ± 0.07^{a}	19.70 ± 0.06^{d}	18.82 ± 0.05^{c}	18.39 ± 0.02^{b}
Crude lipid (%)	whole body	9.37 ± 0.62^{ns}	9.47 ± 1.46	9.53 ± 0.24	7.76 ± 1.80
	liver	29.59 ± 1.04^{b}	26.72 ± 0.96^{ab}	26.53 ± 2.68^{ab}	23.39 ± 2.59^{a}
	dorsal muscle	6.17 ± 0.79^{ns}	5.12 ± 1.93	4.68 ± 0.16	6.53 ± 0.82
Ash (%)	whole body	3.81 ± 0.08^{ns}	3.85 ± 0.03	3.92 ± 0.06	3.83 ± 0.07
	liver	0.65 ± 0.07^{a}	0.76 ± 0.07^{a}	0.95 ± 0.06^{b}	1.10 ± 0.07^{c}
	dorsal muscle	1.30 ± 0.05^{ns}	1.37 ± 0.08	1.34 ± 0.02	1.29 ± 0.24

Data are mean±SD. Values with different superscripts are significantly different ($P<0.05$).
ns not significant.

33

Whole body crude protein was significantly higher in the 10% FSM group than the control, FSM 20% groups ($P < 0.05$). In the livers, the FSM groups were significantly higher than the control group ($P < 0.05$). The dorsal muscles were found to be significantly higher in the FSM groups than the control group. Among FSM groups, FSM 5% was significantly higher than FSM 10% and FSM 20% ($P < 0.05$).

In the crude lipid contents of whole body and dorsal muscle, there were no significant differences between the control and FSM supplement groups. Liver lipid was significantly higher in the control group than the FSM 20% ($P < 0.05$).

No significant differences in the whole bodies and dorsal muscles were found between the control and FSM supplement groups. Liver ash was significantly higher in the FSM 10% and FSM 20% groups than the control and FSM 5% groups ($P < 0.05$).

Table 8. Hematocrit (Ht), Hemoglobin (Hb), glucose, triglyceride, GOT, GPT, total cholesterol, HDL-cholesterol, and total protein of black rock fish, _S. schlegeli_ fed the test diets with various levels of FSM for 8 weeks

	FSM level (%)			
	Control (0)	5	10	20
Ht (%)	36.00 ± 0.89^{ns}	35.83 ± 1.17	34.33 ± 1.97	34.50 ± 2.26
Hb (g/dL)	8.82 ± 1.26^{ns}	8.79 ± 0.62	8.73 ± 0.81	8.54 ± 0.85
Glucose (mg/dL)	4.76 ± 2.37^{ns}	4.56 ± 2.76	3.42 ± 1.24	3.65 ± 2.20
Triglyceride (mg/dL)	408.96 ± 76.02^{ns}	643.88 ± 188.10	481.79 ± 189.42	498.21 ± 16.64
GOT (Karmen/mL)	42.54 ± 26.82^{ns}	71.53 ± 18.61	39.00 ± 18.03	50.43 ± 4.80
GPT (Karmen/mL)	44.29 ± 8.21^{ns}	49.23 ± 7.63	43.03 ± 13.23	45.78 ± 10.63
Total cholesterol (mg/dL)	213.89 ± 67.66^{ns}	289.84 ± 40.48	251.71 ± 50.72	294.98 ± 65.01
HDL-cholesterol (mg/dL)	202.30 ± 26.07^{ns}	210.03 ± 32.18	213.99 ± 34.30	209.84 ± 4.55
Total protein (g/dL)	3.27 ± 0.13^{a}	2.90 ± 0.34^{a}	3.33 ± 0.52^{a}	4.34 ± 0.44^{b}

Data are mean ± standard deviation (SD) of three group of fish (n = 10). Values with different superscripts are significantly different ($P<0.05$). ns not significant

Hematological analysis

Hematocrit (Ht), Hemoglobin (Hb), glucose, GOT, GPT, triglyceride, total cholesterol, HDL-cholesterol, and total protein levels are shown in Table 8. The Ht, Hb, glucose, triglyceride, GOT, GPT, total cholesterol, and HDL-cholesterol levels did not show significant differences between the experimental groups. Total protein was significantly higher in the FSM 20% group than the control, FSM 5%, and FSM 10% groups ($P < 0.05$).

Lysozyme activity in the plasma is shown in Figure 4. The lysozyme activity tended to be higher in FSM supplement groups than the control group, but the differences were not significant.

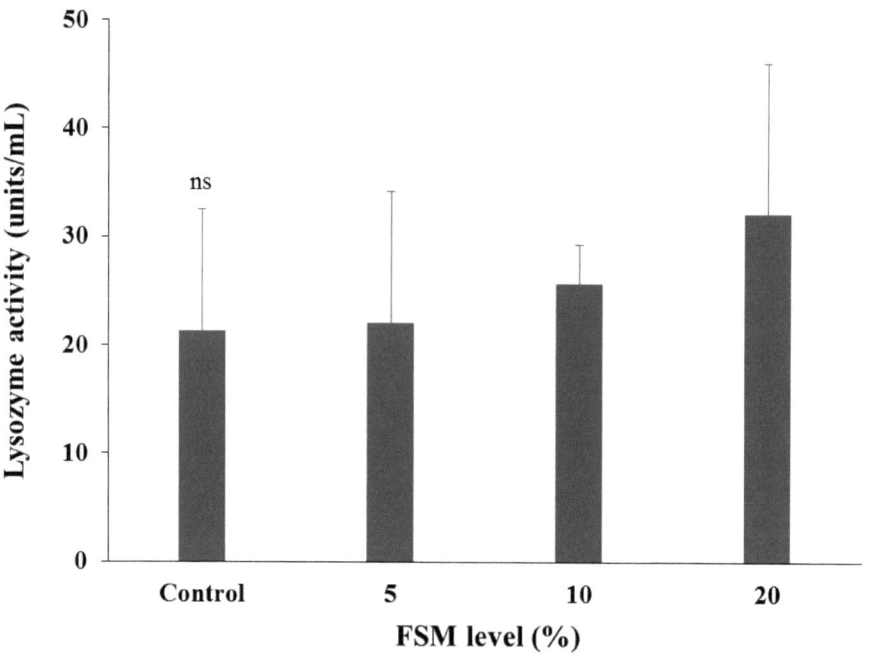

Fig. 4.   Lysozyme activity of black rockfish, *S. schlegeli* fed the test diets with various levels of FSM for 8 weeks. Bar indicates standard deviation (n = 3). ns not significant.

Stress recovery tests

Recovery time after anesthesia with 2-phenoxyethanol is shown in Figure 5. The recovery time after the anesthesia was in a range of 3.5–6.0 min. All FSM supplement groups were significantly faster than the control group ($P < 0.05$), and FSM 20% group had the fastest recovery time.

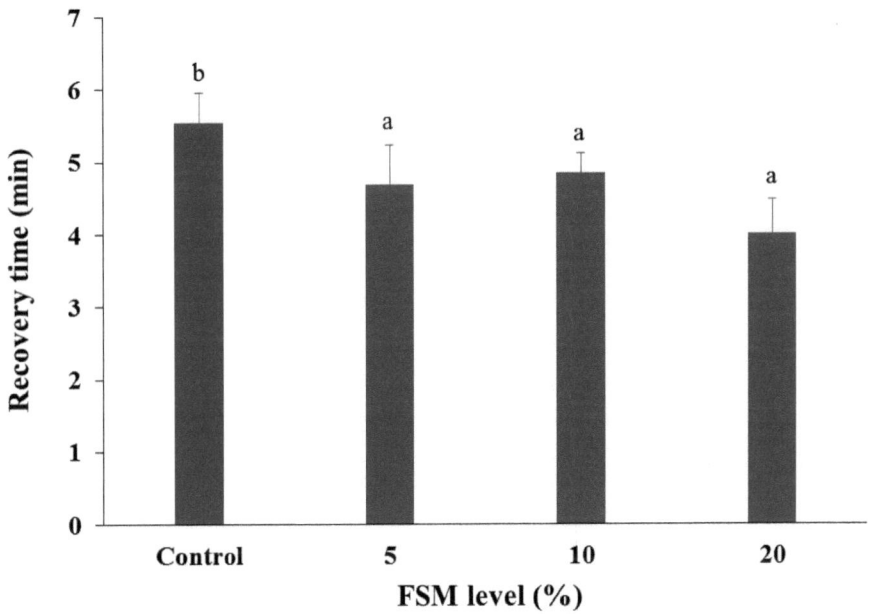

Fig. 5.   Recovery time of black rockfish, *S. schlegeli* fed the test diets with various levels of FSM for 8 weeks after the anesthesia test. Bar indicates standard deviation (n = 3). Bars with different letters differ significantly (P < 0.05).

Mortality after exposure to the air is shown in Figure 6. The mortality was not significantly different in FSM 5% and the control group, whereas it was significantly lower in FSM 10% and FSM 20% compared to the control ($P < 0.05$).

Ht, Hb, glucose, total cholesterol, GOT, and GPT were analyzed during recovery time (Figures 7, 8, 9, 10, 11, 12).

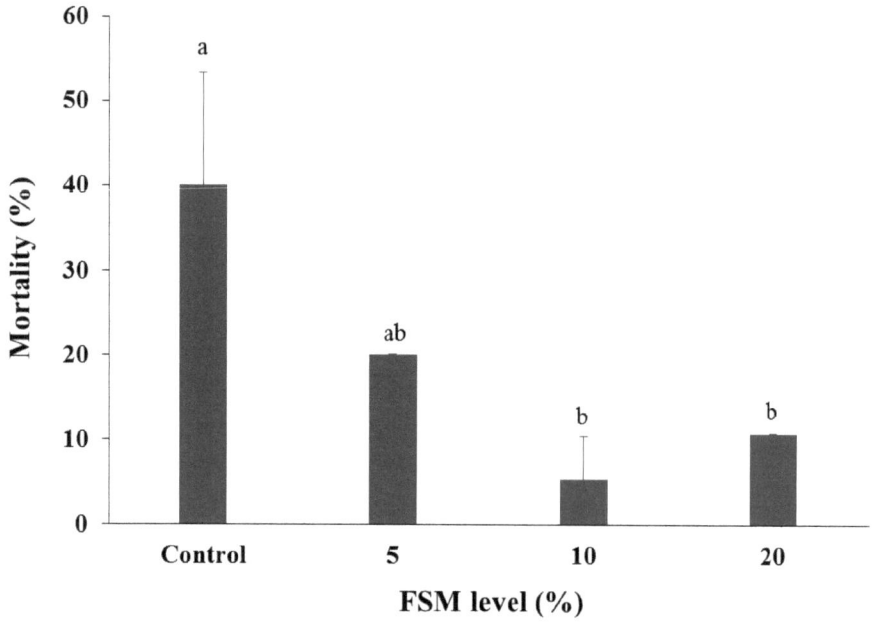

Fig. 6.　Mortality of air exposure test on the black rockfish, *S. schlegeli* fed the test diets with various levels of FSM for 8 weeks. Bar indicates standard deviation (n = 3). Bars with different letters differ significantly (P < 0.05).

At 1 h after the air exposure, the Ht level of FSM 10% was the highest and followed to FSM 5% and FSM 20%, and the control group was the lowest, but the difference was not significant. At 2 h after the air exposure, there were no significant differences observed between the control group and the FSM supplement groups while FSM 5% was significantly higher than FSM 20% ($P < 0.05$). The FSM groups tended to be higher than the control group at 4 h after the exposure, and all Ht of the experimental groups maintained about 40% at 6 h after the exposure, but there were no significant differences either 4 or 6 h after the exposure (Figure 7).

The Hb was not significantly different among the groups at 1 h after the air exposure although it rapidly increased in the control and FSM 5% groups. The control group showed the greatest increase of Hb. At 2 h after the exposure, significant differences were not found among the groups, and the Hb level had remarkably decreased to what it was before the exposure in all groups except for FSM 10% and FSM 20%. At 4 h after the exposure, FSM 20% showed an increase in the Hb level to where it was before the exposure. Significant differences were not observed at 6 h after the exposure and the Hb levels were maintained at the same levels as before the exposure (Figure 8).

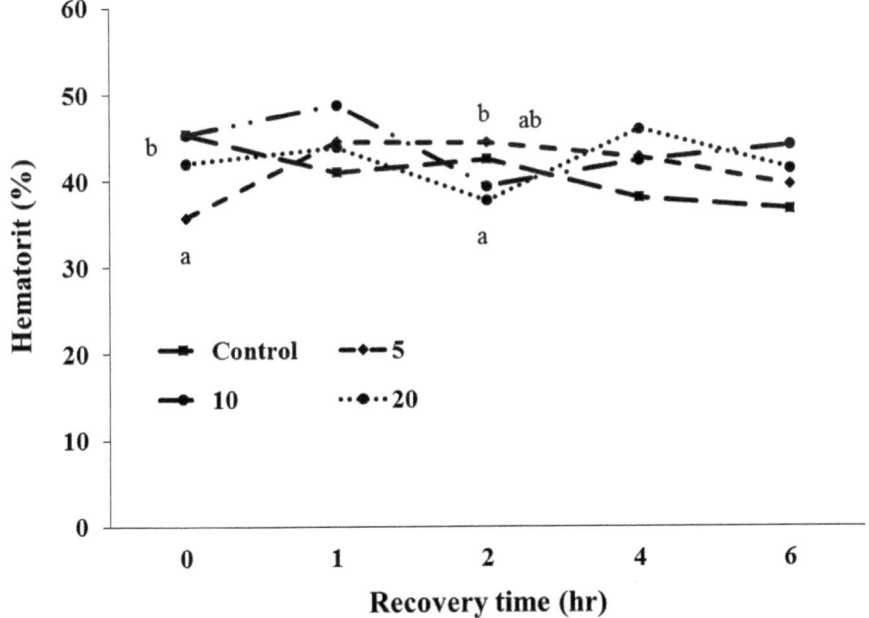

Fig. 7. Changes in plasma hematocrit level of 5-min air exposure test on black rockfish, *S. schlegeli* fed the test diets with various levels of FSM for 8 weeks. Different letters differ significantly (P < 0.05).

At 1 h after the exposure, the glucose levels showed no significant differences among the groups and had remarkably increased in all groups. At 2 h after the exposure, glucose level was significantly higher in the FSM 5% and FSM 10% groups than the control and FSM 20% groups ($P < 0.05$). At 4 h after the exposure, the glucose levels in FSM 10% and FSM 5% were observed to be significantly higher than the control and FSM 20% groups ($P < 0.05$). At 6 h after the exposure, significant differences were present between the groups. However, the glucose level had notably decreased in the FSM 10% and FSM 5% groups and was maintained in control and FSM 20% groups (Figure 9).

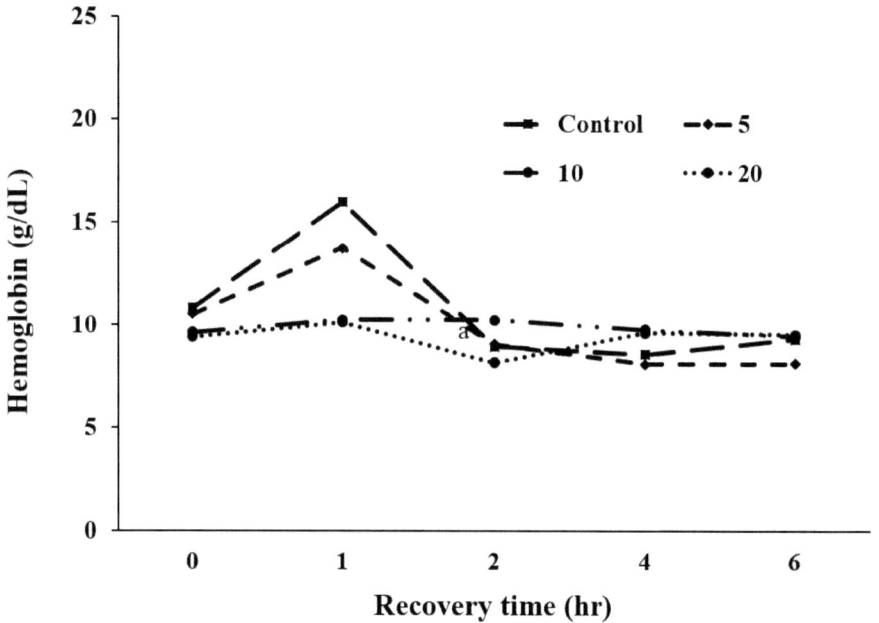

Fig. 8. Changes in plasma hemoglobin level of 5-min air exposure test on black rockfish, *S. schlegeli* fed the test diets with various levels of FSM for 8 weeks. Different letters differ significantly (P < 0.05).

The total cholesterol at 1 h after the exposure was not significantly different among the groups, but had increased in control and decreased in the FSM 5% and FSM 20%. Significant differences were also not observed at 2 h after the exposure, and the total cholesterol tended to have decreased in all groups except for FSM 5%. At 4 h after the exposure, there were no significant differences among the groups, although it had decreased in the control group and FSM 20%, and increased in FSM 10%. At 6 h after the exposure, significant differences were not found between control and FSM supplement groups, while significantly higher total cholesterol was observed in FSM 10% and FSM 20% than the FSM 5% (*P* < *0.05*) (Figure 10).

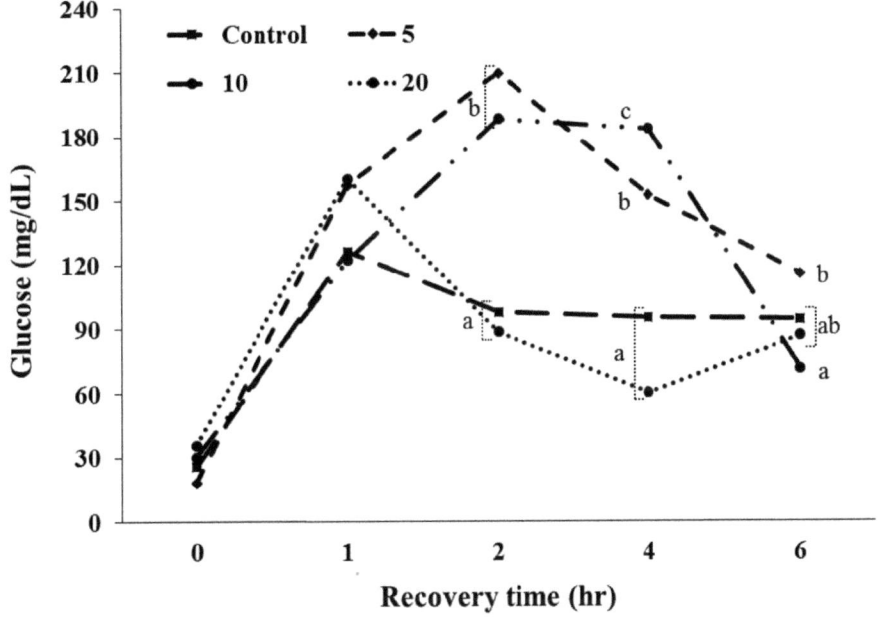

Fig. 9.   Changes in plasma glucose level of 5-min air exposure test on black rockfish, *S. schlegeli* fed the test diets with various levels of FSM for 8 weeks. Different letters differ significantly (P < 0.05).

GOT showed no significant differences among the groups at 1 h after the exposure although it had decreased in all groups. At 2 h after the exposure, FSM 5% had markedly increased, and significantly higher than control, FSM 10%, and FSM 20% groups. At 4 h after the exposure, GOT had decreased in all groups except for FSM 20% and was higher in FSM 5% and FSM 20%. Decreased GOT was observed in all groups at 6 h after the exposure, and significant differences were not observed in all groups (Figure 11).

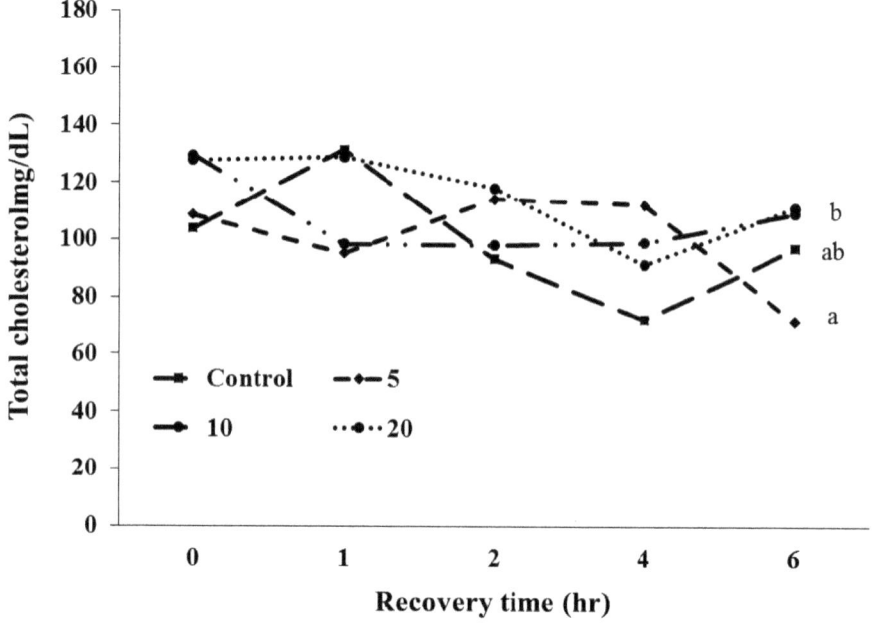

Fig. 10. Changes in plasma total cholesterol level of 5-min air exposure test on black rockfish, *S. schlegeli* fed the test diets with various levels of FSM for 8 weeks. Different letters differ significantly (P < 0.05).

GTP was significantly higher in FSM 5% than the control, FSM 10%, and FSM 20% groups at 1 h after the exposure. At 2 h after the exposure, decreased GTP was found in FSM 5%, and there were no significant differences among the groups. Noticeably increased GPT was observed in all groups except for control at 4 h after the exposure, and FSM supplement group were significantly higher than the control group. GPT increased in FSM 5% but decreased in the other groups at 6 h after the exposure. Especially, GTP was significantly higher in FSM 5% compared to the control group, FSM 10%, and FSM 20% (*P < 0.05*) (Figure 12).

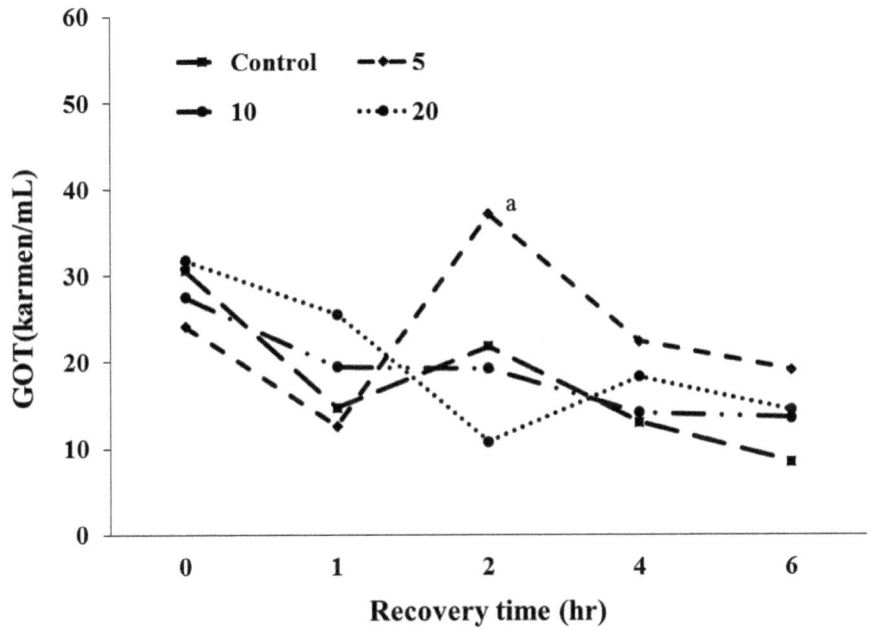

Fig. 11.　Changes in plasma GOT level of 5-min air exposure test on black rockfish, *S. schlegeli* fed the test diets with various levels of FSM for 8 weeks. Different letters differ significantly (P < 0.05).

Discussion

Due to increased prices and unstable supplies of fishmeal, there has been extensive research on alternate protein sources for fish feed (Barrows & Hardy, 2000). Not only have the biochemical properties of these alternative protein sources been studied, but also their feasibility through direct feeding studies in fish (Kim & Bai, 1997; Lee & Yoo, 1996; Lee et al., 2010; Pham et al., 2005; Shapawi et al., 2007; Toppe et al., 2006; Uyan et al., 2006).

In our previous study, biochemical analyses of fish skin from four species of fish (*P. olivaceus*, *S. schlegeli*, *L. maculates*, and *P. major*) were performed (Cho et al.,

2014). The crude protein content of the skin of these fish ranged from 73% to 94% by dry weight; this high level was partly due to a high content of structural protein, collagen. Among the four species, *P. olivaceus* had the thickest dermal and epidermal layers in the dorsal skin. This species was also associated with the highest extraction ratio of acid-soluble collagen. The ASC ratio of the olive flounder skin (20.69%) was highest, followed by that of the red sea bream (20.44%), sea bass (14.74%), and black rockfish (11.00%) based on dry weights. We also examined whether fish skin could be a cost-effective alternative to current fish meal sources. Our analysis indicates that, when it is supplemented with additional fish oils and essential amino acids, fish skin is a viable alternative for fish meal formulations.

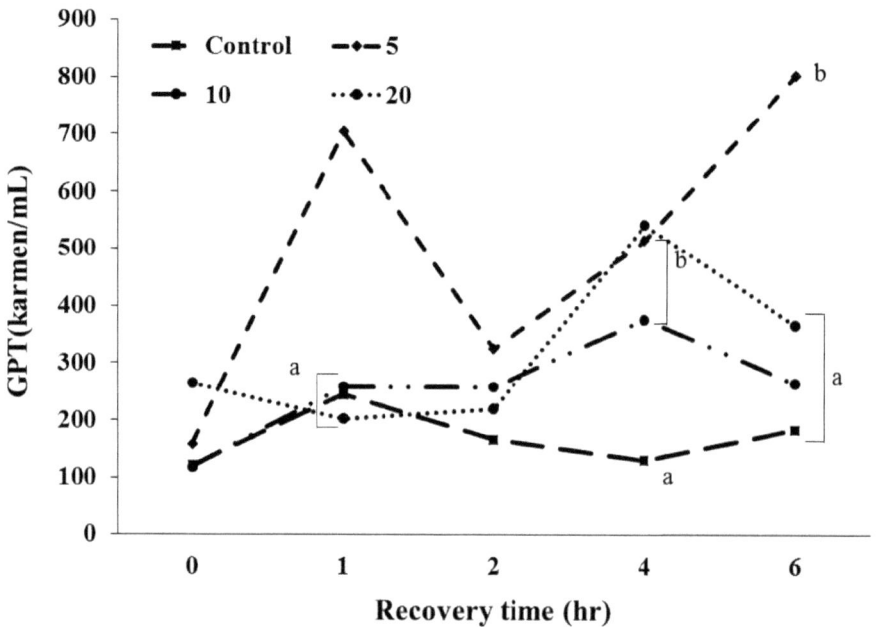

Fig. 12.   Changes in plasma GPT level of 5-min air exposure test on black rockfish, *S. schlegeli* fed the test diets with various levels of FSM for 8 weeks. Different letters differ significantly (P < 0.05).

According to Kim and Bai (1997), when a fishmeal mixed with animal protein sources including blood meal, squid liver meal, meat and bone meal, greaves, feather meal, and essential amino acids that were gradually substituted for to 20–60% of the regular diet was fed to black rockfish for 6 weeks, growth performance did not significantly change at concentrations up to 40% of the fishmeal. In addition, once black rockfish were fed the fishmeal mixed with 12.5–50% animal protein supplement for 16 weeks, up to 12.5% of the fishmeal could be substituted before the rate of gain and feed efficiency started showing large differences as the level of supplement increased. This was reported to be because of diet intake preferences, *in vivo* enzyme activity, and metabolism (Kim & Bai, 1999). In the present study, although significant differences were not found between the control group and the FSM supplement groups, it tended to decrease gradually with an increase of the supplement. Such an FSM supplement at the same fishmeal ratio seems to be capable of facilitating the preference and metabolism of fish, and vitamin C addition would provide synergetic effects in the FSM supplement. Aksnes et al (2008) added 0.7–5.6 g/kg of hydroxyproline to vegetable protein diets and fed them to salmon (*Salmo salar* L.) for 88 days. They found that adding 2.9 g/kg of hydroxyproline to the diet increased the growth of the salmon by approximately 14%. The diet from the current study is also expected to improve the growth performance because the FSM used in this study contains abundant hydroxyproline, which is a collagen-specific amino acid.

Meanwhile, the condition factor (CF), viscerasomatic index (VSI), and hepatosomatic index (HSI) were significantly lower in the FSM supplement groups. In particular, CF was observed to be significantly lower in the FSM supplement groups compared to the control group. When black rockfish were fed a diet with fishmeal supplement mixed with animal proteins, Kim and Bai (1997, 1999) observed that HSI increased in 2.8 g juvenile fish with an increase in the supplement, and HIS and CF decreased in 21.1 g juvenile fish, which indicates discrepant size-dependent results, even in the same kind of fish. However, when using soybean meal as an alternative fishmeal, decreased HIS and CF were found in juvenile *Paralichthys olivaceus* in a dose-dependent manner (Kim et al., 2000b). HIS was not significantly different in *Hexagrammos otakii* Jordan et Starks when soybean meal and feather meal were added gradually as fishmeal replacement (Lee & Lee, 1998). Meanwhile, hybrid striped bass (*Morone chrysops* ♀ × *M. saxatilis* ♂) fed diets with some added poultry by-products as protein sources to replace fishmeal showed significantly increased HIS and intraperitoneal fat (IPF) when up to 70% of the fishmeal was substituted (Rawles et al., 2006). Additionally, high fat content in the fish body was found in silver seabream (*Rhabdosargus sarba*) and Nile tilapia (*Oreochromis niloticus* L.) with the addition of poultry by-products (El-Sayed, 1994; 1998). It was also reported that body fat and HIS (Steffens, 1994) and VSI and HSI (Zoccarato et al., 1996) increased in *Oncorhynchus mykiss* in response to poultry by-products supplementation. Comparing the results of such fishmeal substitution is difficult because of the various protein sources and fish utilized. However, the nutritional characteristics of the alternative protein sources affect lipid metabolism in the body such that HSI, CF, and VSI would be influenced. In particular, as FSM contains relatively higher lipids and proteins compared to fishmeal, it would affect lipid accumulation and metabolism. In the FSM supplement group, the moisture and protein content in the whole body increased and the lipid content decreased, and the ash content decreased significantly compared to the control group. Such results were in agreement with the moisture and protein content in muscles, and the liver showed the same results with the protein and ash content in the whole body. However, the lipids in the liver were confirmed to be higher than the control group, indicating that lipid accumulation in the liver increased. The body composition of fish is affected by various factors such as intraspecific strain differences, water temperature, and increased body weight, and is influenced the most by the amount of feed supplied and the mix proportions of the feed (Nandeesha et al., 1995; Zeitler et al., 1984). It was also reported that lipids and moisture in fish bodies decreased with fish growth while protein and mineral content mostly remained unchanged (Murai et al., 1985). Kim and Bai (1997, 1999) reported that feeding black rockfish with an animal protein supplement had no effects on body composition unless there were nutritional problems, which differs from this study. However, Bai et al (1998) found that when *Cyprinus carpio* were fed diets with added greaves and meat and bone meal containing over 2 times more crude fat than fishmeal, the crude fat accumulated in the body was used as an energy source, thereby increasing protein accumulation. This is in agreement with the present study that added FSM (17%) possessing over 2 times higher lipid content compared to white and brown fishmeal (7–8%). Even though black rockfish and *Cyprinus carpio* live in different habitats and the composition of the greaves, meat and bone meal, and FSM are different, the lipid content in the FSM seems to be used as energy sources because both fishes use lipids as an energy source. This was partially confirmed by the increased crude protein content and the decreased lipid content in the whole fish and muscles. Rogie and Skinner (1985) found that lipoproteins of rainbow trout were synthesized in the liver cells, and Lee et al (2000) mentioned that inhibited lipoprotein synthesis owing to the impaired metabolism of liver cells could result in lipid accumulation in the liver without using them in the liver in black rockfish. In the current study, HIS was decreased with respect to the FSM supplementation, but the lipids in the liver were increased and tended to accumulate. It is difficult to conclude whether such results affect lipoprotein metabolism due to the FSM supplement in the feed. However, based upon the high lipid content in the liver of the control group, there seems to be relationships between FSM supplementation and vitamin C addition. Further studies need to be done on this topic.

This study observed no significant differences in hematocrit, hemoglobin, glucose, GOT, GPT, total cholesterol, or HDL-cholesterol in the blood components. The lysozyme activity was not significantly different among the FSM supplement groups and the control group. However, the triglycerides tended to decrease, and the total protein in FSM 20% was found to be significantly higher. Changes in the plasma constituents are critical indicators of the health status of organisms since they can differ due to a lack of essential nutrients in feed (Murai et al., 1982a) or the feeding environment of the fish (Park et al., 1999). In general, hemoglobin levels of healthy fish are known to be 9–10 g/dl (Post, 1983). Hemoglobin and hematocrit levels of 21.4 g black rockfish fed a fishmeal replacement were reported to be 7.3–9.0 g/dl and 40.2–45.4%, respectively. These were slightly lower and higher, respectively, than the hemoglobin and hematocrit levels in the present study, indicating that the adequate amount for fish is still not clear (Kim & Bai, 1999).

Lysozymes, lymphocyte-derived mucous bacteriolytic enzymes with antibiotic-like characteristics, are widely distributed in nature and have been found in secretions such as mucous, saliva, and blood. Lysozymes dissolve acetyl-monopolysaccharides in the cell walls of gram-positive bacteria and show direct bacteriolytic action in gram-negative bacteria. In accordance with the complement and antibody supports, the bacteriolytic actions are also improved, which are complement-dependent. Therefore, lysozymes are closely related to the complement and become a part of the intrinsic defense mechanism against parasitic, bacterial, and viral infections in a number of animals (Ingram, 1980; Lim et al., 2009). In this study, lysozymes did not show significant differences among the FSM supplement groups and the control group. However, lysozymes tended to increase, so it is possible that they are affected by the FSM. Vitamin C increases resistance to diseases in fish (Verlhac & Gabauda, 1994; 1997). Therefore, synergetic effects with the FSM supplement are also expected in the presence of long-term feeding.

47

There are various factors that induce stress in fish. In particular, the handling, sorting, shipment, and transportation necessary during fish farming and bad water quality stress fish (Barton & Iwama, 1991; Schreck, 1982; Wendelaar Bonga, 1997). In fish farming and the fishery industry, various harvesting methods (Harman et al., 1980; Hopkins & Cech, 1992; Maule & Mesa, 1994; Maule et al., 1989; Mitton & McDonald, 1994), physiological shock (Gadomski et al., 1994; Sverdrup et al., 1994), and exposure to contaminated environments (Marshall Adams, 1990; Brown, 1993; Cairns et al., 1984; Folmar, 1993; Kime, 1995; Niimi, 1990) induce stress (Iwama et al., 1997). Stress affects every part of the fish from community structures to biochemical instability (Marshall Adams, 1990) and is also a cause of mass mortality in fish farming (Iwama et al., 1997). In the present study, the FSM supplement groups showed fast recovery in the anesthesia and air exposure tests. Ji et al (2007) observed fast recovery after anesthesia and positive effects regarding the air exposure in juvenile *Paralichthys olivaceus* fed diets supplemented with medicinal herbs for 8 weeks. This was suggested to be because of improvements in liver functions and active biosynthesis of glycogen owing to the addition of herbs with antioxidant properties. Similar results were also confirmed in juvenile *Pagrus major* fed medicinal herbs for 12 weeks. This was also ascribed to the medicinal herbs containing antioxidants that prevent oxidative damages inducing aging, functional impairments, and diseases (Ji et al., 2009). The skin of *Paralichthys olivaceus* used in this study is a protein source containing lots of collagen, but no antioxidant functions of collagen have been reported. To study the prevention of collagen destruction, collagen from squid skin was added to human dermal cells, cultured for 24 h, and analyzed through UV for 24 h (Kwon et al., 2008). This treatment also accelerated differentiation by facilitating the growth of cell lines cultured with type I collagen (Kim et al., 2000a). In the human body, melanin, which raises pigmentation such as spots, freckles, and age spots in the presence of excess production, thereby promoting skin damage, is produced during the biosynthesis of tyrosinase (Invergar & McEvily, 1992), and a study confirmed tyrosinase inhibition in the type 1 collagen of calves and the collagen extracted from squid skin (Kwon et al., 2008). This tyrosinase inhibitory activity is associated with phenol content and antioxidant effects (Ra et al., 1997), and the small phenols in collagen were reported to be the major cause of tyrosinase inhibition (Kwon et al., 2008) in a study that did not mention antioxidant effects. Matsuda et al (2006) reported that skin fibroblast density and collagen density significantly increased in young pigs fed collagen peptides. Because the previous studies were mostly performed at the cell level or with terrestrial animals, it is difficult to make direct comparisons. However, the factors related to collagen are expected to have positive effects on cell aging, especially the antioxidant properties. According to Walrand et al (2008), when dairy products with added collagen hydrolysates were given to 15 men, the concentration of blood amino acids related to collagen greatly increased and collagen synthesis was facilitated. Even through the results were obtained from humans, collagen is considered to influence blood biochemical factors and energy metabolism as this study also found a significant decrease in blood lipids and an increase in blood total proteins. However, since there are no studies on collagen intake in aquatic organisms like fish, systematic studies regarding the intake of fish and factors related to collagen are necessary in near future.

Schreck (1982) mentioned that blood lactic acid, lipids, electrolytes, hemoglobin (Hb), total protein, hematocrit (Ht), and liver glycogen could be used to determine the presence or absence of stress. In addition to the stress indicators, such items are indicators used in the evaluation of health in fish bodies (Wedemeyer & Mcleay, 1981; Wedemeyer & Yasutake, 1977). Following the air exposure test in this study, stress responses included a rapid increase in hemoglobin (Hb) 1 h after exposure. Significant differences were not observed among the experimental groups, although the increase in the FSM groups was smaller than that in the control group. Generally,

Ht and Hb are known to be indicators of oxygen transportation in the body (Min et al., 2003). Stress in marine fish was reported to increase blood Ht and Hb levels (Davis & Parker, 1990). In this study, however, Ht did not change considerably, and Hb tended to increase. Such a result is considered to be due to the differences in the response time of Hb rather than Ht. Further, as the air exposure test was directly related to oxygen consumption, oxygen was rapidly consumed within 2 h. The recovery rate seems to have differed due to the rapid supplementation of hemoglobin, especially in the FSM groups compared to the control group.

Generally, blood glucose concentrations are elevated by stress (Olsen et al., 1995). Barton and Iwama (1991) mentioned that increased glucose levels with an increase in cortisol levels were the result of a second reaction owing to stress hormones. Ji et al (2009) reported that the rapid recovery of blood cortisol and glucose to the levels from before the air exposure test was observed in the medicinal herb supplement group, which showed significant growth when juvenile *Pagrus major* fed with medicinal herb supplemented diets were exposed to the air. This was suggested to be because of the rapid reduction of consumption of the stored energy owning to stress. In the present study, we also observed an increase in the glucose concentration in all of the experimental groups due to drastic energy consumption owing to stress, and rapid recovery tended to be exhibited in the groups with high weight gain, which was in agreement with the results of Ji et al (2009). The stress is capable of inhibiting growth by consuming energy stored in the body due to liver gluconeogenesis and increased aminotransferase activity as well as protein catabolism (Davis et al., 1985).

The total cholesterol also tended to be decreased in all experimental groups after the air exposure, which indicates that energy consumption due to increased metabolism occurs in lipids.

The amine transferases GOT and GPT are distributed in blood in the liver and spleen of fish bodies. While they maintain low activity when the fish body is healthy, they are released in the presence of tissue necrosis or diseases, resulting in increased activity (Min et al., 2003). GOT was observed to increase in trout when stress was induced with rapid changes of the salinity in the feeding water (Chang and Hur, 1999). Juvenile black sea breams also showed increased GOT with acute salt stress (Min et al., 2003). Significantly increased GOT was found in adult and juvenile *Paralichthys olivaceus* in response to acute stress from low salt levels in the feeding water (Her et al., 2002). Her et al (2002) reported that this was due to impaired liver functions owing to physiological burdens on liver and spleen cells and excess energy consumption. However, although we observed a decrease in GOT and an increase in GPT, results that differed from previous studies, we were able to confirm that the stress due to air exposure negatively influenced liver functions. The FSM groups had especially high GPT levels compared to the control group. Even though such a difference is considered to mainly be caused by vitamin C concentrations, it showed higher values in the supplement group with greater than FSM10% than the control group. It is therefore expected that FSM functional substances such as collagen have positive effects that promote the quick recover of energy metabolism following impaired liver functions.

In this study, the black rockfish was fed a diet in which 5%, 10%, or 20% of the casein was substituted with FSM with vitamin C levels of 400 mg/kg for 8 weeks. Significantly higher growth performance and feeding efficiency were observed in FSM 20% compared to the casein supplement group. CF, HSI, and VSI levels and lipid content in the body were lower while protein accumulation increased. Essential amino acids in the body also tended to increase, and amino acids and free amino acids showed different patterns to each other and confirmed that FSM supplementation influenced amino acid composition. There were also differences observed in organic acids and free sugars. Blood composition, lysozyme activity, and stress responses were investigated to confirm the health status of the fish, and it was found that stress decreased blood lipids and increased blood proteins. Significant differences were not found in the lysozyme activity, but it tended to increase with FSM supplementation. In the air exposure and anesthesia tests, the stress response was significantly higher in the FSM supplement group, and the FSM groups also showed rapid recovery after air exposure. Taken together, FSM are superior protein sources for the black rockfish compared to purified proteins like casein, and health improvements in the farmed fish are also expected. However, the discrepant results of the vitamin C supplementation indicate that detailed studies are required regarding the relationship between FSM and vitamin C.

Chapter 3

Nutritional characteristics of black rockfish (*Sebastes schlegeli*) fed a diet of fish skin

This study investigated the effects of diets substituted with different levels (0, 5, 10, and 20 %) of flounder skin meal (FSM) on the nutritional composition of black rockfish *Sebastes schlegeli*. Fish (10.05 ± 0.44 g) were fed to apparent satiation twice daily for 8 weeks. Adding FSM decreased crude lipid levels and increased crude protein and ash. The abundant fatty acids in the FSM-added group were C16:0, C18:1-cis (n9), and C22:6n-3. The major amino acids in the samples were glutamic acid, aspartic acid, glycine, leucine, alanine, lysine, and arginine. The abundant free amino acids in the FSM-added group were taurine, glutamic acid, alanine, leucine, and arginine. Six free sugars were found in all groups. Glucose was predominant, followed by mannose, rhamnose, fucose, fructose, and ribose. Among the three organic acids in the whole body of black rockfish, lactic acid was predominant, followed by citric acid and oxalic acid. Total organic acid content in the control was significantly higher than those of FSM substitution groups.

Introduction

Numerous studies have been investigated vegetable and animal proteins that could replace fishmeal in fish feed.   In particular, there have been many studies on the use of vegetable protein sources such as soybean meal (Kim et al., 2009; Lee & Jeon, 1996; Murai et al., 1982; Robert et al., 1993), cottonseed meal, and rapeseed meal (Lee & Yoo, 1996; Pham et al., 2005), which have a relatively stable supply compared to fishmeal, to replace fishmeal as a source of protein.   However, plant resources are constantly in competition with livestock and human consumption, and the recent development of plant extract fuels such as bioethanol will eventually lead to an increase in the price of the plant resources usable as protein resources (Rena & Hasan, 2009).   The by-products of the processing of terrestrial livestock such as cows, chickens, and pigs could be used as animal protein sources, since they have a relatively high protein content and qualitatively similar amino acid composition to fishmeal, and are inexpensive and stably supplied. Various studies have been conducted on their use as protein sources to replace fishmeal in fish feed (Ai et al., 2006; Bai et al., 1998; Kikuchi et al., 1994; Kim & Bai, 1999; Lee & Lee, 1998; Sato & Kikuchi, 1997). However, the rise of safety issues due to serious infectious diseases like mad cow disease, swine fever, and avian influenza has gradually restricted the use of livestock by-products lately. Thus, as there are economic and safety issues with using terrestrial protein sources to replace fishmeal, securing economic and safe protein sources from marine products rather than terrestrial products is necessary. Many researchers have investigated by-products obtained from processing marine animal as potential protein sources, including shrimp by-products (Cruz-Suárez et al., 1993), tuna muscle by-products (Uyan et al., 2006), shrimp and fish by-products (Li et al., 2004), squid liver meal mixing soybean meal with by-products of squid processing (Lee & Lee, 1998), fish bone and crab by-products (Goytortúa-Bores et al., 2006; Lee et al., 2010; Toppe et al., 2006), and fish by-products (Foster et al., 2005). Of the fishery by-products, even though fish skins obtained from the consumption of raw fish are a good protein source because of high collagen content, by-products such as bones and internal organs are only partially used and mostly discarded.

Therefore, this study was conducted to investigate improvement of quality and physiological function on cultured black rockfish fed diets substituted with different levels (0, 5, 10, and 20 %) of flounder skin meal (FSM).

Materials and Methods

The skin of *Paralichthys olivaceus*, which has the highest farming yield and raw fish consumption in Korea, is easy to secure in large quantities due to its low use, thickness, and high collagen content, and was obtained from nearby fish markets. The fish skin was washed with fresh water and was subjected to hot air drying (50-60°C) followed by grinding via a high speed grinder (ZM-1000, Retsch Co., Japan) to prepare flounder skin meal (FSM).

Table 9. Ingredient and proximate composition of experimental diets with various levels of FSM.

Ingredient	g/100g			
	Control (0)	5	10	20
White fish meal	41	41	41	41
Casein	20	15	10	-
Flounder skin meal (FSM)	-	5	10	20
L-ascorbic acid	0.02	0.02	0.02	0.02
Wheat flour	22.58	23.28	23.98	25.28
Feed oil (squid liver oil)	8.4	7.7	7	5.7
α-potato starch	3	3	3	2
Vitamin premix[1] (vitamin C free)	2	2	2	2
Mineral Premix[2]	2	2	2	1
Choline Chloride	1	1	1	1
Total	100	100	100	100
Vitamin C in diets	212.41	197.16	204.67	203.12
*Proximate analysis				
Protein	46.71	47.21	47.11	46.11
Lipid	9.21	9.51	10.11	10.81
Ash	8.21	8.61	8.71	8.71

[1]Vitamin premix (mg/g mixture) : retinol acetate, 0.81mg; cholecalciferol, 0.012mg; vitamin E, 22.5mg; vitamin K_3, 2.5mg, thiamine, 5.5mg; riboflavin, 10mg; pyridoxine, 6mg; niacin, 37.5mg; folic acid, 2mg; biotin 0.05mg; inositol 50mg. All ingredients were diluted with alpha-cellulose to 1g.

[2]Mineral premix (mg/g mixture) : Mn, 3.2mg; Zn, 3.2mg; Fe, 3.0mg; Cu, 0.36mg; $MgSO_4$, 100mg; KCl (47%), 60mg; $Al(OH)_3$, 1.06mg; $Ca(IO_3)_2$, 0.475mg; $CoSO_4$, 0.475mg. All ingredients were diluted with alpha-cellulose to 1g.

*Dry matter

After acclimation for 2 months in a square stock tank (running water system, 6.0 m × 6.0 m

× 1.2 m) at the Fisheries Science Institute, Chonnam National University, Korea., 45 juvenile fish (mean body weight, 10.05 ± 0.44 g) were randomly selected from the stock tank and transferred to separate 300-L rectangular tanks (running water system, 1.0 m × 0.8 m × 0.8 m). The flow rate of filtered seawater in each tank was adjusted to 5 L/min. Mean water temperature, salinity, and dissolved oxygen were 20.2 ± 2.3 °C, 32.0 ± 1.2 psu, and 6.3 ± 0.4 mg/L, respectively, and were measured using a YSI-85 (YSI, Ohio, USA) probe. The rearing trial was conducted in triplicate for each tested diet. The fish were fed twice a day (at 0800 h and 1600 h), until apparent satiation, for 8 weeks. The amount of feed given to each tank was recorded daily to calculate feeding efficiency.

Ingredients and proximate compositions of the experimental diets in response to FSM substitution and the results of vitamin C analyses are shown in Table 9. Proximate analyses were carried out to evaluate the nutritional composition of the prepared diets, and the vitamin C content of the diets were analyzed using the 2,4-dinitrophenyl hydrazine (DNP) colorimetric method (Al-Ani et al., 2007).

FSM is a high protein meal containing more than 80% crude proteins but is lacking in essential amino acids compared to fishmeal.   When fishmeal is replaced by FSM, unknown factors present in essential amino acids and fishmeal may affect the experimental fish, making it difficult to evaluate the influences of FSM substitution on experimental fish.   Therefore, to maintain proper balances of the essential amino acids and to minimize the effects of the unknown factors in the fishmeal, white fishmeal (FF Skagen LT Supreme, Denmark) was fixed at the same level throughout the experimental diets.   Casein, a purified protein, was used to control the protein content of each experimental diet.   Squid liver oil (Ihwa, Korea) rich in DHA and EPA, which are essential fatty acids for the black rockfish, was used as a lipid source. Wheat flours (CJ, Korea) and α-potato starch were employed as carbohydrate sources to control energy and bind the diets.   To find the relationship between the level of vitamin C (Sandel & Daniel, 1988) and FSM substitution on the fish body, vitamin C (200 mg/kg) was added based upon the vitamin C requirements of the black rockfish reported by Bai et al (1996).

There were a total of 4 experimental groups, including a control group with fishmeal and casein only and 3 experimental groups with 5, 10, or 20% of the casein replaced by FSM.

Analyzes of proximate composition, fatty acid, total amino acid, free amino acid, organic acid, and free sugar in this study were carried out based on AOAC methods (2002) and some modifications of Hwang et al (2010).

All mean values were analyzed via one-way analysis of variance (ANOVA). When differences were found among data, Duncan's multiple range test was used to compare the mean difference by using the SPSS software package version 17 (Statistical Package for Social Sciences, SPSS Inc., Chicago, IL, USA). Differences were considered significant at $p < 0.05$.

Results

Table 10. Proximate composition (%) of whole body in black rockfish (*S. schlegeli*) fed the test diets with various levels of FSM for 8 weeks.

Proximate composition (g/100g)	FSM substitution level (%)			
	Control (0)	5	10	20
Moisture	68.14±1.28[a]	67.93±0.47[a]	68.73±0.05[ab]	69.54±0.07[b]
Crude protein	14.28±0.07[a]	15.10±0.03[d]	14.75±0.08[b]	14.93±0.04[c]
Crude lipid	10.53±0.22[c]	9.72±0.02[b]	9.20±0.25[b]	8.25±0.53[a]
Ash	3.77±0.07[a]	4.11±0.08[c]	3.92±0.08[b]	4.04±0.03[bc]

Data are mean±SD. Values with different superscripts are significantly different (*P<0.05*).
ns not significant.

Table 11. Fatty acid composition of whole body in black rockfish (*S. schlegeli*) fed the test diets with various levels of FSM for 8 weeks (g/100g).

Fatty acid	FSM substitution level (%)			
	Control (0)	5	10	20
C12:0	0.31±0.01[d]	0.26±0.01[c]	0.24±0.01[b]	0.19±0.00[a]
C13:0	0.05±0.01[ns]	0.05±0.00	0.05±0.00	0.05±.00
C14:0	7.42±0.05[b]	6.91±0.23[a]	7.14±0.17[ab]	7.33±0.07[b]
C15:0	0.96±0.07[a]	1.00±0.02[ab]	1.00±0.01[ab]	1.04±0.02[b]
C16:0	24.07±0.06[b]	23.23±0.46[a]	23.69±0.22[ab]	23.44±0.15[a]
C17:0	0.79±0.01[bc]	0.79±0.00[c]	0.75±0.03[ab]	0.72±0.02[a]
C18:0	6.69±0.10[b]	6.81±0.13[b]	6.63±0.16[b]	6.24±0.07[a]
C20:0	1.64±0.04[c]	1.03±0.03[b]	0.96±0.02[a]	0.92±0.05[a]
C21:0	0.57±0.02[ns]	0.56±0.11	0.55±0.01	0.50±0.05
C22:0	0.81±0.01[a]	0.81±0.01[a]	0.76±0.01[b]	0.72±0.01[a]
C23:0	1.12±0.02[a]	1.25±0.06[b]	1.34±0.03[c]	1.55±0.03[d]
C24:0	1.43±0.02[a]	1.50±0.01[c]	1.45±0.01[ab]	1.47±0.03[bc]
Saturates	45.87±0.23[b]	44.21±0.60[a]	44.58±0.21[a]	44.18±0.06[a]
C14:1	0.32±0.01[d]	0.27±0.01[a]	0.30±0.01[b]	0.31±0.00[c]
C15:1	0.02±0.00[ns]	0.02±0.00	0.02±0.00	0.02±0.00
C16:1	6.29±0.03[b]	5.99±0.10[a]	6.33±0.02[b]	6.46±0.07[c]
C17:1	0.45±0.01[a]	0.45±0.01[a]	0.46±0.00[a]	0.49±0.01[b]
C18:1n9t	0.53±0.19[ns]	0.37±0.17	0.46±.12	0.39±0.10
C18:1n9c	20.43±0.15[b]	20.45±0.25[b]	20.40±0.08[b]	19.89±0.13[a]
C20:1	4.65±0.11[ab]	5.15±0.34[b]	4.72±0.51[ab]	4.23±0.20[a]
C22:1n9	0.96±0.01[a]	0.97±0.01[ab]	0.99±0.01[bc]	0.99±0.01[c]
C24:1	0.89±0.11[ns]	1.06±0.17	1.04±0.17	1.14±0.16
Monoenes	34.55±0.28[ns]	34.73±0.52	34.71±0.61	33.92±0.49
C18:2n6t	0.17±0.01[ns]	0.17±0.01	0.17±0.01	0.12±0.08
C18:2n6c	0.43±0.01[ns]	0.42±0.01	0.43±0.01	0.43±0.01
C20:2	0.77±0.00[a]	0.83±0.02[b]	0.78±0.02[a]	0.76±0.02[a]
C22:2	0.07±0.01[a]	0.11±0.01[b]	0.10±0.02[ab]	0.09±0.02[ab]
C18:3n6	0.33±0.01[a]	0.36±0.01[b]	0.33±.01[a]	0.35±0.01[b]
C18:3n3	1.31±0.01[ns]	1.32±0.04	1.30±0.02	1.30±0.01
C20:3n6	0.03±0.01[ns]	0.02±0.00	0.02±0.00	0.03±0.01
C20:3n3	0.15±0.01[ns]	0.17±0.04	0.15±0.01	0.16±0.02
C20:4n6	0.26±0.00[a]	0.28±0.01[b]	0.29±0.00[c]	0.31±0.00[d]
C20:5n3	0.00±0.00[a]	0.28±0.08[c]	0.19±0.02[b]	0.18±0.02[b]
C22:6n3	16.04±0.47[a]	17.09±0.87[ab]	16.95±0.42[a]	18.17±0.40[b]
Polyenes	19.57±0.47[a]	21.06±0.87[b]	20.71±0.41[b]	21.91±0.47[c]
n3	17.51±0.48[a]	18.87±0.88[bc]	18.59±0.45[ab]	19.82±0.43[c]
n6	1.22±0.01[ns]	1.25±0.03	1.24±0.02	1.24±0.08
n3/n6	14.31±0.33[a]	15.07±1.00[ab]	15.04±0.52[ab]	16.06±0.73[b]

Data are mean±SD. Values with different superscripts are significantly different (*P<0.05*). ns not significant.

Proximate compositions with various levels of FSM are shown in Table 10. Crude protein was significantly higher in the FSM groups than the control group. FSM 5% was much higher than FSM 10% and FSM 20% (P < 0.05). Crude lipid was significantly lower in the FSM groups than the control group, especially in FSM 20% compared to that of FSM 5% and FSM 10% (P < 0.05).

The fatty acid composition of the whole body is shown in Table 11. Saturates were observed to be significantly lower in all FSM groups than the control group (P < 0.05), and no significant differences were found in monoenes (P < 0.05). Significantly higher polyenes were observed in the FSM groups than the control group, especially in FSM 20% (P < 0.05). The control group was also significantly lower in n-3 than FSM 5% and FSM 20% (P < 0.05), whereas FSM 10% was not significantly different from the control group. There were no significant differences in n-6 between the experimental groups. The n-3/n-6 ratio was not significantly different between the control group and FSM 5% and FSM 10%, while FSM 20% was significantly higher than the control group (P < 0.05).

The whole-body amino acid contents are shown in Table 12. The total amino acids and EAA were significantly higher in FSM 20% than the other groups (P < 0.05). FSM 20% was significantly higher in all amino acids compared to the control group (P < 0.05), whereas FSM 5% and FSM 10% were not significantly different from the control group.

The whole-body free amino acid contents are shown in Table 13. The total free amino acids were significantly lower in the FSM groups than the control group, and the FSM groups tended to decreased significantly in a dose-dependent manner (P < 0.05).

Seven kinds of free sugars were analyzed, and fucose, rhamnose, glucose, mannose, fructose, and ribose were detected but not galactose (Table 14). The total free sugars were not significantly different in the control group, FSM 5%, and FSM 10%, while they were significantly lower in FSM 20% (P < 0.05).

Six kinds of organic acids were analyzed, and lactic acid, oxalic acid, and citric acid were found, but not malic acid, tartaric acid, or maleic acid (Table 15). The total organic acid content was significantly lower in the FSM groups than the control group (P < 0.05).

Table 12. Total amino acid content of whole body in black rockfish (*S. schlegeli*) fed the test diets with various levels of FSM for 8 weeks (g/100g)

	FSM substitution level (%)			
	Control (0)	5	10	20
Aspatic acid	2.97±0.07[a]	3.20±0.30[a]	3.17±0.04[a]	3.58±0.12[b]
*Threonine	1.54±0.05[a]	1.66±0.16[ab]	1.73±0.21[ab]	1.82±0.04[b]
Serine	1.85±0.06[a]	2.02±0.19[ab]	2.01±0.10[ab]	2.23±0.07[b]
Glutamic acid	3.78±0.09[a]	4.12±0.39[a]	4.08±0.15[a]	4.60±0.16[b]
Proline	1.59±0.03[a]	1.73±0.17[a]	1.67±0.06[a]	1.97±0.09[b]
Glycine	4.37±0.11[a]	4.89±0.45[a]	4.58±0.11[a]	5.62±0.33[b]
Alanine	3.11±0.09[a]	3.41±0.31[a]	3.18±0.10[a]	3.79±0.17[b]
Cystine	0.11±0.00[a]	0.12±0.03[ab]	0.13±0.01[ab]	0.15±0.01[b]
*Valine	1.61±0.05[a]	1.74±0.18[ab]	1.74±0.03[ab]	1.93±0.06[b]
*Methionine	0.81±0.02[a]	0.86±0.10[a]	0.86±0.01[a]	0.97±0.03[b]
*Isoleucine	1.23±0.02[a]	1.32±0.13[a]	1.31±0.01[a]	1.46±0.04[b]
*Leucine	2.22±0.05[a]	2.37±0.23[a]	2.34±0.01[a]	2.60±0.07[b]
*Tyrosine	0.57±0.03[a]	0.61±0.11[a]	0.67±0.04[a]	0.76±0.01[b]
*Phenylalanine	0.96±0.02[a]	1.07±0.11[a]	1.04±0.05[a]	1.19±0.04[b]
*Histidine	0.81±0.02[a]	0.91±0.07[b]	0.86±0.01b[a]	0.93±0.02[b]
*Lysine	1.72±0.04[a]	1.86±0.21[a]	1.94±0.03[a]	2.25±0.08[b]
Ammonia	2.28±0.05[a]	2.42±0.17[ab]	2.32±0.07[a]	2.58±0.11[b]
*Arginine	1.42±0.04[a]	1.48±0.18[a]	1.52±0.02[a]	1.76±0.07[b]
Total	32.96±0.82[a]	35.80±3.44[a]	35.17±0.76[a]	40.18±1.49[b]
*EAA	12.32±0.31[a]	13.28±1.35[a]	13.35±0.34[a]	14.90±0.45[b]

Data are mean±SD. Values with different superscripts are significantly different (*P<0.05*).

ns not significant.

*Essential amino acid.

Table 13. Free amino acid content of whole body in black rockfish (*S. schlegeli*) fed the test diets with various levels of FSM for 8 weeks (g/100g)

	FSM substitution level (%)			
	Control (0)	5	10	20
Phosphoserine	0.77 ± 0.03^b	0.84 ± 0.01^c	0.78 ± 0.02^b	0.62 ± 0.02^a
Taurine	3.31 ± 0.04^a	3.27 ± 0.17^a	3.41 ± 0.04^a	3.57 ± 0.09^b
Aspartic acid	3.31 ± 0.02^d	2.51 ± 0.02^c	1.61 ± 0.04^b	1.35 ± 0.02^a
Hydroxyproline	1.68 ± 0.05^b	1.23 ± 0.27^a	1.89 ± 0.13^b	1.70 ± 0.18^b
Threonine	3.18 ± 0.04^a	2.26 ± 0.06^b	1.54 ± 0.05^c	1.13 ± 0.02^d
Serine	2.19 ± 0.02^d	1.84 ± 0.09^c	1.39 ± 0.04^b	0.91 ± 0.02^a
Asparagine	0.88 ± 0.06^b	0.30 ± 0.02^a	0.30 ± 0.01^a	0.31 ± 0.00^a
Glutamic acid	5.46 ± 0.10^d	3.91 ± 0.05^c	2.57 ± 0.07^b	2.36 ± 0.05^a
Proline	1.95 ± 0.04^d	1.51 ± 0.10^c	1.19 ± 0.03^b	0.75 ± 0.03^a
Glycine	2.75 ± 0.03^c	2.97 ± 0.11^d	2.57 ± 0.09^b	2.38 ± 0.01^a
Alanine	5.95 ± 0.05^d	4.92 ± 0.08^c	3.79 ± 0.13^b	3.02 ± 0.01^a
Citrulline	0.34 ± 0.03^d	0.22 ± 0.04^c	0.11 ± 0.01^b	0.05 ± 0.00^a
Valine	3.10 ± 0.02^d	2.07 ± 0.05^c	1.53 ± 0.05^b	0.99 ± 0.02^a
Methionine	0.86 ± 0.02^d	0.48 ± 0.01^c	0.41 ± 0.02^b	0.32 ± 0.02^a
Isoleucine	2.46 ± 0.01^d	1.62 ± 0.02^c	1.10 ± 0.04^b	0.73 ± 0.01^a
Leucine	7.33 ± 0.01^d	4.33 ± 0.03^c	3.34 ± 0.13^b	2.54 ± 0.01^a
Tyrosine	1.35 ± 0.06^b	1.62 ± 0.06^c	1.44 ± 0.06^b	0.99 ± 0.03^a
Phenylalanine	2.84 ± 0.01^d	1.61 ± 0.02^c	1.33 ± 0.03^b	0.95 ± 0.01^a
β-aminoisobutyric acid	1.06 ± 0.07^b	0.53 ± 0.02^a	0.61 ± 0.12^a	0.55 ± 0.03^a
γ-amino-n-butyric acid	0.03 ± 0.01^b	0.01 ± 0.00^a	0.01 ± 0.00^a	0.02 ± 0.00^{ab}
Histidine	0.95 ± 0.01^d	0.56 ± 0.02^c	0.45 ± 0.02^b	0.20 ± 0.01^a
1-methylhistidine	0.06 ± 0.00^b	0.03 ± 0.01^a	0.02 ± 0.00^a	0.01 ± 0.01^a
Carnosine	0.21 ± 0.01^c	0.13 ± 0.03^a	0.14 ± 0.01^a	0.07 ± 0.02^a
Anserine	0.29 ± 0.06^{ns}	0.29 ± 0.05	0.31 ± 0.07	0.32 ± 0.01
Tryptopan	0.51 ± 0.09^b	0.22 ± 0.08^a	0.17 ± 0.07^a	0.09 ± 0.04^a
Hydroxylysine	0.11 ± 0.01^b	0.05 ± 0.01^a	0.04 ± 0.01^a	0.05 ± 0.02^a
Ornitine	0.19 ± 0.01^c	0.09 ± 0.01^b	0.06 ± 0.01^a	0.06 ± 0.00^a
Lysine	2.99 ± 0.07^d	0.77 ± 0.03^c	0.49 ± 0.02^b	0.33 ± 0.02^a
Ammonia	2.20 ± 0.03^c	1.86 ± 0.16^a	1.42 ± 0.03^a	1.55 ± 0.02^b
Ethanolamine	0.36 ± 0.08^{ns}	0.40 ± 0.03	0.34 ± 0.03	0.38 ± 0.04
Arginine	8.71 ± 0.12^d	5.84 ± 0.18^c	5.25 ± 0.16^b	3.57 ± 0.05^a
Total	67.37 ± 0.26^d	48.31 ± 0.99^c	39.61 ± 1.29^b	31.88 ± 0.22^a

Data are mean±SD. Values with different superscripts are significantly different (*P<0.05*).
ns not significant.

Table 14. Free sugar content of whole body in black rockfish (*S. schlegeli*) fed the test diets with various levels of FSM for 8 weeks (mg/L)

	FSM substitution level (%)			
	Control (0)	5	10	20
Fucose	150.20 ± 15.39^a	155.82 ± 10.22	132.53 ± 14.89	145.44 ± 15.64
Rhamnose	354.54 ± 35.57^d	284.62 ± 27.03^c	136.03 ± 32.66^b	82.35 ± 0.55^a
Glucose	280.99 ± 5.25^a	424.50 ± 31.76^b	464.12 ± 27.22^b	289.80 ± 19.75^a
Mannose	190.91 ± 11.39^a	217.63 ± 32.91^a	319.56 ± 14.43^b	207.09 ± 17.14^a
Fructose	76.05 ± 8.64^a	87.12 ± 2.18^{ab}	105.61 ± 7.69^{bc}	109.35 ± 15.86^c
Ribose	81.90 ± 6.05^c	75.70 ± 7.02^{bc}	65.63 ± 4.73^b	49.10 ± 7.70^a
Total	$1,134.59\pm82.29^b$	$1,245.38\pm111.11^b$	$1,223.49\pm101.61^b$	883.14 ± 76.63^a

Data are mean±SD. Values with different superscripts are significantly different (*P<0.05*).
ns not significant.

Table 15. Organic acid content of whole body in black rockfish (*S. schlegeli*) fed the test diets with various levels of FSM for 8 weeks (mg/L)

	FSM substitution level (%)			
	Control (0)	5	10	20
Lactic acid	12.12 ± 0.34^c	10.11 ± 0.09^b	7.79 ± 1.63^a	9.09 ± 0.13^{ab}
Oxalic acid	0.19 ± 0.01^b	0.15 ± 0.02^a	0.13 ± 0.01^a	0.15 ± 0.03^a
Citric acid	0.72 ± 0.03^a	0.68 ± 0.02^{ab}	0.65 ± 0.03^a	0.71 ± 0.02^b
Total	13.03 ± 0.12^c	10.90 ± 0.18^b	8.57 ± 1.60^a	9.95 ± 0.16^{ab}

Data are mean±SD. Values with different superscripts are significantly different (P<0.05).
ns not significant.

Discussion

In the previous result, it was confirmed that there was a high collagen content (approximately 20% dry weight) in *Paralichthys olivaceus* skin (Cho et al., 2014). Flounder skin meal (FSM) replaced various fractions of the casein substitution in the fishmeal, and this was fed to the fish for 8 weeks to confirm the substitution effect.

Comparing the results of such fishmeal substitution is difficult because of the various protein sources and fish utilized. However, the nutritional characteristics of the alternative protein sources affect lipid metabolism in the body. In particular, as FSM contains relatively higher lipid content (17%) compared to white and brown fishmeal (7-8%), it would affect lipid accumulation and metabolism. In the FSM substitution group, the protein content in the whole body increased and the lipid content decreased, and the ash content increased significantly compared to the control group. The body composition of fish is affected by various factors such as intraspecific strain differences, water temperature, and increased body weight, and is influenced the most by the amount of feed supplied and the mix proportions of the feed (Nandeesha et al., 1995; Zeitler et al., 1984).

Among whole-body fatty acid responses to the FSM substitution in the feed, saturated fatty acids significantly decreased while polyunsaturated fatty acids increased, especially n-3 HUFA, which is an essential fatty acid in black rockfish (Lee et al., 1993). The white fishmeal used in this study contained about 0.7% n-3 HUFA, and fish oil, like the squid liver oil used as a lipid source, has more than 20% n-3 HUFA with an appropriate EPA/DHA ratio (Kalogeropoulos et al., 1992). Accumulation of n-3 HUFA in fish bodies increased because of the FSM substitution. This seems to be because of FSM-specific amino acids rather than the differences in n-3 HUFA in the feed.

Amino acids in fish were significantly higher only in FSM 20% than the control group. Although such differences may occur due to the FSM substitution, this is not certain because significant differences in the amino acids were not observed between the control group and the experimental groups. However, the free amino acids in the fish body were significantly decreased in the FSM groups compared to the control group, suggesting that the FSM substitution influenced amino acid metabolism in the fish. In marine animals, free amino acids provide chemical signals for behaviors, communication, and metabolism through sensory organs (Saglio et al., 1990). Moreover, free amino acids act as substrates for protein biosynthesis or aerobic catabolism and provide osmolality stably during embryonic stages through intrinsic nutrients in marine fish (Ronnestad & Fyhn, 1993; Sivaloganathan et al., 1998).

Lactic acids were the most abundant organic acids, especially in the FSM substitution groups. Lactic acids are known to greatly differ based on the amount of activity at the time of harvesting and the storage conditions (Park et al., 1997). However, lactic acids showed significant differences in the FSM substitution group compared to the control group in the present study. Given that the amount of activity at the time of harvesting and storage conditions were similar in this study, lactic acids seem to have an influence on the energy metabolism of glycolysis with respect to the FSM substitution.

The glucose content in the free sugars was significantly higher in FSM 5% and FSM 10% than the control group; but no significant differences between control and FSM 20%, indicating that FSM substitution affect glucose metabolism, resulting in differences in body glucose. Ribose differed in a similar manner, leading to glucose levels decreased in response to the FSM substitution. Free ribose is abundant in the muscles of living fish but is also released from inosine, which is produced by ATP decomposition after death. Consequently, ribose content can depend on the pretreatment conditions of the samples immediately after the instant killing (Kim et al., 2009). As ribose, along with fructose, is a quantitatively important factor in glycolysis, it is considered to be somewhat associated with glucose metabolism.

Acknowledgments

This research was supported by Basic Science Research Program through the National Research Foundation of Korea (NRF) funded by the Ministry of Education, Science and Technology (2011-0011204) and the Program for the Construction of Eco Industrial Park (EIP) which was conducted by the Korea Industrial Complex Corporation (KICOX) and the Ministry of Trade, Industry & Energy (MOTIE).

References

Ai Q, Mai K, Tan B, Xu W, Duan Q, Ma H, Zhang L (2006) Replacement of fish meal by meat and bone meal in diets for large yellow croaker, *Pseudosciaena crocea*. Aquaculture 260: 255-263.

Aksnes A, Mundheim H, Toppe J, Albrektsen S (2008) The effect of dietary hydroxyproline supplementation on salmon (*Salmo salar* L.) fed high plant protein diets. Aquaculture 275:242–249

Al-Ani M, Opara LU, Al-Bahri D, Al-Rahbi N (2007) Spectrophotometric quantification of ascorbic acid contents of fruit and vegetables using the 2,4-dinitrophenylhydrazine method, Journal of Food, Agriculture and Environment 5: 165-168.

AOAC (2002) Official methods of analysis, 16th edn. Association of Official Analytical Chemists, Arlington.

AOCS (1990). *Official Methods and Recommended Practice of the AOCS* (4th ed.). Illinois: The American Oil Chemists Society.

Arvanitoyannis IS, Kassaveti A (2008) Fish industry waste: treatments, environmental impacts, current and potential uses. International Journal of Food Science & Technology, 43, 726-745.

Bai SC, Daniel KJ, Jang HK (1996) Development and Experimental Model for Vitamin C Requirement Study in Korean Rockfish, *Sebastes schlegeli*. Journal of Aquaculture 9: 169-178.

Bai SC, Jang HK, Cho ES (1998) Possible Use of the Animal By-product Mixture as a Dietary Fish meal Replace in Growing Common Carp (*Cyprinus carpio*). Korean Journal of Fisheries Society 31: 380-385.

Bai SC, Lee KJ, Jang HK (1996) Development of an experimental model for vitamin C requirement study in Korean rockfish, *Sebastes schlegeli*. Journal of Aquaculture 9:169-178.

Barrows FT, Hardy RW (2000) Encyclopedia of aquaculture. John Wiley & Sons, New York,

USA, 1063pp.

Barton BA, Iwama GK (1991) Physiological changes in fish from stress in aquaculture with emphasis on the response and effects of corticosteroids. Annual Review of Fish Disease 1:3–26.

Bechtel PJ (ed.) (2003) Advances in Seafood Byproducts: 2002 Conference Proceedings. Alaska, 515pp.

Bechtel PJ, Smiley S (2010) A Sustainable Future: Fish Processing Byproducts. Alaska Sea Grant, University of Alaska Fairbanks, 340pp.

Brown JA (1993) Endocrine responses to environmental pollutants. In: Rankin JC, Jensen FB (eds) Fisheries ecophysiology. Chapman and Hall, London, UK, pp 276–296.

Cairns VW, Hodson PV, Nraigu JO (1984) Contaminant effects on fisheries. John Wiley and Sons, New York, NY, USA, 333pp.

Chalamaiah M, Dinesh Kumar B, Hemalatha R, Jyothirmayi T (2012) Fish protein hydrolysates: Proximate composition, amino acid composition, antioxidant activities and applications: A review. Food Chemistry, 135, 3020-3038.

Chan D, Lamande SR, Cole WC, Bateman JF (1990) Regulation of procollagen synthesis and processing during ascorbate-induced extracellular matrix accumulation in vitro. Biochemical Journal, 269: 175–181.

Chang YJ, Hur JW (1999) Physiological responses of grey mullet (*Mugil cephalus*) and Nile tilapia (*Oreochromis niloticus*) by rapid changes in salinity of rearing water. Journal of Korean Fisheries Society 32:310–316.

Cho JK, Jin YG, Rha SJ, Kim SJ, Hwang JH (2014) Biochemical characteristics of four marine fish skins in Korea. Food Chemistry 159:200–207.

Cruz-Suárez LE, Ricque-Marie D, Martínez-Vega JA, Wesche-Fbeling P (1993) Evaluation of two shrimp by-product meals as protein sources in diets for *Penaeus vannamei*. Aquaculture 115:53–62.

Davis KB, Parker NC (1990) Physiological stress in striped bass: effect of acclimation temperature. Aquaculture 91:349–358.

Davis KB, Torrance P, Parker NC, Suttle MA (1985) Growth, body composition and hepatic tyrosine aminotransferase activity in cortisol-fed channel catfish, *Ictalurus punctatus* Rafinesque. Journal of fish biology 27:177–184.

Dong FM, Hardy RW, Haard NF, Barrows FT, Rasco BA, Falrgrleve WT, Forster LP (1993). Chemical composition and protein digestibility of poultry by-product meals from salmonid diets. Aquaculture, 116, 149-158.

Drury RB, Wallington EA (1980) *Carleton's Histological Technique*. (5th ed.). Oxford: Oxford University Press, 520pp.

Duncan DB (1955) Multiple range and multiple F tests. Biometrics 11:1–42.

El-Sayed AFM (1994) Evaluation of soybean meal, spirulina meal and chicken offal meal as protein sources for silver seabream (*Rhabdosargus sarba*) fingerlings. Aquaculture 127:169–176.

El-Sayed AFM (1998) Total replacement of fish meal with animal protein sources in Nile tilapia, *Oreochromis niloticus* (L.), feeds. Aquaculture Research 29: 275–280.

Eyre DR, Wu JJ (1987) Type XI or $1\alpha2\alpha3\alpha$ collagen. In: Mayne R, Burgeson RE (eds) Structure and function of collagen types. Academic Press, New York, USA, pp 261–281.

Fairbanks G, Steck TL, Wallach DF (1971) Electrophoretic analysis of the major polypeptides of the human erythorocyte membrane. Biochemistry, 10, 2606-2616.

Fessler JH, Fessler LI (1987) Type V collagen. In: Mayne R, Burgeson RE (eds) Structure and function of collagen types. Academic Press, New York, USA, pp 81–103.

Folmar LC (1993) Effects of chemical contaminants on blood chemistry of teleost fish: a bibliograhy and synopsis of selected effects. Environmental Toxicology and Chemistry 12: 337–375.

Foster I, Babbitt JK, Smiley S (2005) Comparison of the nutritional quality of fish meals made from by-products of the Alaska fishing industry in diets for Pacific threadfin (*Polydactylus sexfilis*). Journal of World Aquaculture Society 36: 530–537.

Gadomski DM, Mesa MG, Olson TM (1994) Vulnerability to predation and physiological stress responses of experimentally descaled juvenile chinook salmon, *Oncorhynchus tshawytscha*. Environmental Biology of Fishes 39: 191–199.

Goytortúa-Bores E, Civera-Cerecedo R, Rocha-Meza S, Green-Yee A (2006) Partial replacement of red crab (*Pleuroncodes planipes*) meal for fish meal in practical diets for the white shrimp *Litopenaeus vannamei*. Effects on growth and in vivo digestibility. Aquaculture 256:414–422.

Goytortúa-Bores E, Civera-Cerecedo R, Rocha-Meza S, Green-Yee A (2006) Partial replacement of red crab (*Pleuroncodes planipes*) meal for fish meal in practical diets for the white shrimp *Litopenaeus vannamei*. Effects on growth and in vivo digestibility. Aquaculture 414-422.

Harman BJ, Johnson DL, Greenwald L (1980) Physiological responses of Lake Erie freshwater drum to capture by commercial shore seine. Transactions of the American Fisheris Society 109:544–551.

Her JW, Lee BK, Chang YJ, Lee JK, Lim YS, Lee JH, Park CH, Kim BK (2002) Stress responses of olive flounder *Paralichthys olivaceus* to hyposalinity. Journal of Aquaculture 15:69–75.

Hopkins TE, Cech JJJ (1992) Physiological effects of capturing striped bass in gill nets and fyke traps. Transactions of the American Fisheris Society 121:819–822.

Hwang J, Mizuta S, Yokoyama Y, Yoshinaka R (2007). Purification and characterization of molecular species of collagen in the skin of skate (*Raja kenojei*). Food Chemistry, *100*, 921–925.

Hwang JH, Lee SW, Rha SJ, Jeong DH, Han KH, Shin TS (2010) Nutritional Characteristics of Eels (*Auguilla japonica*) Fed a Diet of Yuza (*Citrus junos* Sieb ex Tanaka). Korean Journal of Fisheries and Aquatic Sciences 43: 573-580.

Ingram GA (1980) Substances involved in the natural resistance of fish to infection-a review. Journal of Fish Biology 16:23–60.

Invergar R, McEvily AJ (1992) Studies on biological activity from extract of *Crataegi fructus*. Korean Journal of Herbology 17:29–38.

Iwama GK, Pickering AD, Sumpter JP, Schreck CB (1997) Stress in finfish: past, present and future - a historical perspective. In: Iwama GK, Pickering AD, Sumpter JP (eds) Fish stress and health in aquaculture. Cambridge University Press, London, UK, pp 1–34.

Jayathilakan, K., Sultana, K., Radhakrishna, K., Bawa, A.S. (2012). Utilization of byproducts and waste materials from meat, poultry and fish processing industries: a review. Journal of Food Science and Technology, 49, 278-293.

Ji SC, Jeong GS, Im GS, Lee SW, Yoo JH, Takii K (2007) Dietary medicinal herbs improve growth performance, fatty acid utilization, and stress recovery of Japanese flounder. Fish Science 73:70–76.

Ji SC, Takaoka O, Lee SW, Hwang JH, Kim YS, Ishimaru K, Seoka M, Jeong GS, Takii K (2009) Effect of dietary medicinal herbs on lipid metabolism and stress recovery in red seabream, *Pagrus major*. Fish Science 75:665–672.

Kalogeropoulos N, Alexis MN, Henderson JJ (1992) Effect of dietary soybean and cod-liver oil levels on growth on growth and body composition of gilthead bream (*Sparus aurata*). Aquaculture 104: 293-308.

Kikuchi K, Honda H, Kiyono M (1994) Utilization of feather meal also a protein source in the diet of juvenile Japanese flounder. Fish Science 60: 203-306.

Kim JS (1996) Quality improvement of surimi gel from fish with a red muscle by emulsion curd containing a modified fish skin gelatin. Journal of the Korean Society for Agriculture, Chemistry, Biotechnology 39:361–367.

Kim HY, Kim EH, Kim DH, Oh MJ, Shin TS (2009) The Nutritional Components of Olive Flounder (*Paralichthys olivaceus*) Fed Diets with Yuza (*Citrus junos* Sieb ex Tnaka). Korean Journal of Fisheries and Aquatic Sciences 42: 215-223.

Kim KW, Bai SC (1997) Fish meal analog as a dietary protein source in Korean rockfish, *Sebastes schlegeli*. Journal of Aquaculture 10:143–151.

Kim KW, Bai SC (1999) Possible use of the dietary fish meal analogue in juvenile Korean rockfish, *Sebastes schlegeli*. Journal of Korean Fisheries Society 32:149–154.

Kim JS, Cho SY (1996) Screening for raw material of modified gelatin in marine animal skins caught in coastal offshore water in Korea. Journal of the Korean Society for Agriculture, Chemistry, Biotechnology 39:134–139.

Kim JS, Kim JG, Cho SY, Kang KS, Ha JH, Lee EH (1993b) The suitable processing condition for gelatin preparation from yellowfin sole skin. Korean Journal of Food Science and Technology 25:716–723.

Kim JS, Ihm CW, Kim PH (1996) Preparation and properties of gelatin from conger eel skin. Agric Chem Biotechnol 39:274–281.

Kim JS, Kim JG, Cho SY (1997) Screening for the raw material of gelatin from the skins of some pelagic fishes and squid. Journal of Korean Fisheries Society 30:55–61.

Kim JS, Kim JD, Kang MJ, Ahn HY, Kim DJ (2000a) Collagen-induced activation of MMPs (membrane-type matrix metalloproteinase and matrix metalloproteinase-2) in ovarian cancer cell lines in vitro. Journal of Korean Society of Obstetrics and Gynecology 43:1972–1978.

Kim SG, Kang OJ, Kwak DC (1993a) Physicochemical characteristics of filefish and cod skin collagen. Journal of Korean Agricultural Chemistry Society 36:163–171.

Kim YS, Kim BS, Moon TS, Lee SM (2000b) Utilization of defatted soybean meal as a substitute for fish meal in the diet of juvenile flounder (*Paralichthys olivaceus*). Journal of Korean Fisheries Society 33:469–474.

Kim SG, Kwak DC (1991) The enzymatic modification and functionalities of filefish skin collagen. Journal of Korean Agricultural Chemistry Society 34:265–272.

Kim, K.W., Park, G.J., Ok, I.H., Bai, S.C., Choi, Y.J., & Shin, I.S. (2002). Effects of dietary synthetic amino acid supplementation in Korean rockfish fry, *Sebastes schlegeli*. Journal of Aquaculture, 15, 157-163.

Kim MH (2004) The effect of L-ascorbic acid on the oxidative reaction of lysine in collagen.

Journal of Life Science 14:478–483.

Kim MH (2006) Effect of L-ascorbic acid on collagen synthesis in 3 T6 fibroblasts and primary cultured cells of chondrocytes. Journal of Korean Society of Food Science and Nutrition 35:42–47.

Kime DE (1995) The effects of pollution on reproduction in fish. Reviews in Fish Biology and Fisheries 5:52–95.

Kimura S (1997) Collagen types of fish. In: Kimura S (ed) The extracellular matrix of fish and marine invertebrates. Kouseisha-Kouseikaku, Tokyo, Japan, pp 9–17.

Kuhn K (1987) The classical collagens : type I, II and III collagen. In: Mayne R, Burgeson RE (eds) Structure and function of collagen types. Academic Press, New York, USA, pp 1–42.

Kwon MC, Qadir SA, Kim HS, Ahn JH, Cho NH, Lee HY (2008) UV protection and whitening effects of collagen isolated from outer layer of the squid *Todarodes pacificus*. Journal of Korean Fisheries Society 41:7–12.

Laemmli, U.K. (1970). Cleavage of structural protein during the assembly of the head of bacteriophage T4. *Nature, 227*, 680-685.

Lee JK, Lee SM (1998) Evaluation of soybean meal or feather meal as a partial substitute for fish meal in formulated diets for fat cod (*Hexagrammos otakii* Jordan at Starks). Journal of Aquaculture 11:421–428.

Lee JK, Lee SM (1998) Evaluation of Soybean Meal or Feather Meal as a Partial Substitute for Fish Meal in Formulated Diets for Fat Cod (*Hexagrammos otakii* Jordan at Starks). Journal of Aquaculture 11: 421-428.

Lee KJ, Powell MS, Barrows FT, Smiley S, Bechtel P, Hardy RW (2010) Evaluation of supplemental fish bone meal made from Alaska seafood processing byproducts and dicalcium phosphate in plant protein based diets for rainbow trout (*Oncorhynchus mykiss*). Aquaculture 302:248–255.

Lee SM, Jeon IG (1996) Evaluation of soybean meal as a partial substitute for fish meal in formulated diets for Korean rockfish, *Sebastes schlegeli*. Journal of Korean Fisheries Society 29:586–594.

Lee SM, Kim DJ, Kim JK, Hhr SB, Lee JK, Lim HK (2000) Effects of *Kluyveromyces fragilis*, *Candida Utilis* and brewer's yeast as an additive in the diet on the growth and body composition of juvenile Korean rockfish (*Sebastes schlegeli*). Korean Fisheries Society 33:463–468.

Lee SM, Lee JY (1996) Substitution of plant and animal proteins for fish meal in the practical formulated feeds for juvenile Korean rockfish (Sebastes schlegeli). Korean Journal of Animal Nutrition & Feedstuffs, 20, 409-418.

Lee SM, Lee JY, Kang YJ, Hur SB (1993) Effects of Dietary n-3 Highly Unsaturated Fatty Acid on Growth and Biochemical Changes in the Korean Rockfish *Sebastes schlegeli*. I. Growth and Body Composition. Journal of Aquaculture 6: 89-105.

Lee SM, Yoo JH (1996) Evaluation of Cotton Seed Meal or Rapeseed Meal as a Partial Substitute for Fish Meal in Fotmulated Diets for Korean Rockfish (*Sebastes schlegeli*). Korean Journal of Animal Nutrition & Feedstuffs 20: 128-135.

Lee SM, Yoo JH (1996) Evaluation of cotton seed meal or rapeseed meal as a partial substitute for fish meal in formulated diets for Korean rockfish (*Sebastes schlegeli*). Korean Journal of Animal Nutrition and Feed 20:128–135.

Li P, Wang X, Hardy RW, Gatlin DM III (2004) Nutritional value of fisheries by-catch and by-product meals in the diet of red drum (*Sciaenops ocellatus*). Aquaculture 236:485–496.

Lim DK, Yoo KY, Shin DG, Kim JE, Bae JY, Bai SC, Lee JY (2009) Effects of dietary kugija *Lycium chinense* supplementation on juvenile Korean rockfish *Sebastes schlegeli*. Korean Journal of Fisheries and Aquatic Sciences 42:250–256.

Lupi O (2002) Prions in dermatology. Journal of the American Academy of Dermatology 46:790–793.

Marshall Adams S (1990) Biological indicators of stress in fish. American Fisheries Society, pp. 191.

Matsuda N, Koyama Y, Hosaka Y, Ueda Y, Watanabe T, Araya T, Irie S, Takehana K (2006) Effects of ingestion of collagen peptide on collagen fibrils and glycosaminoglycans in the dermis. Journal of Nutritional Science and Vitaminology 52:211–215.

Maule AG, Mesa MG (1994) Efficacy of electrofishing to assess plasma cortisol concentration in juvenile chinook salmon passing hydroelectric dams on the Columbia river. North American Journal of Fisheries Management 14:334–339.

Maule AG, Tripp RA, Kaattari SL, Schreck CB (1989) Stress alters immune function and disease resistance in chinook salmon (*Oncorhynchus tshawytscha*). Jouranl of Endocrinology 120:135–142.

Min BH, Kim BK, Hur JW, Bang IC, Byun SK, Choi CY, Chang YJ (2003) Physiological responses during freshwater acclimation of seawater-cultured black porgy (*Acanthopagrus schlegeli*). Korean Journal of Ichthyology 15:224–231.

Mitton CJ, McDonald DG (1994) Consequences of pulsed DC electrofishing and air exposure to rainbow trout (*Oncorhynchus mykiss*). Canadian Journal of Fisheries and Aquatic Sciences 51:1791–1798.

Murai T, Akiyama T, Hirasawa Y, Oshiro T, Okauchi M, Nose T (1982a) Blood constituent levels and body composition of wild and cultured bluefin tuna juveniles. Bulletin of National Resarch Institute of Aquaculture 3:51–59.

Murai T, Ogata H, Nose T (1982b) Methionine coated with various materials supplemented to soybean meal diet for fingerling carp *Cyprinus carpio* and channel catfish *Lctalurus punctatus*. Bulletin of Japanese Society of Scientific Fisheries 48:85–88.

Murai T, Akiyama T, Watanabe T, Nose T (1985) Effects of dietary protein and lipid levels on performance and carcas composition of fingerling carp. Nippon Suisan Gakgaishi 51:605–608.

Muyonga, J.H., Cole, C.G.B., & Duodu K.G. (2004). Characterization of acid soluble collagen from skins of young and adult Nile perch (*Lates niloticus*). Food Chemistry, 85, 81-89.

Nandeesha MC, De Silva SS, Murthy DS (1995) Use of mixed feeding schedules in fish culture: performance of common carp, *Cyprinus carpio* L., on plant and animal protein based diets. Aquaculture Research 26:161–166.

Niimi AJ (1990) Review of biochemical methods and other indicators to assess fish health in aquatic ecosystems containing toxic chemicals. Journal of Great Lakes Research 16:529–541.

Nurad S, Sivarajah A, Pinnel SR (1981) Regulation of propyl and lysyl hydroxylase activities in cultured human skin fibroblasts by ascorbic acid. Bichemical and Biophysical Research Communications 101:868–875.

NRC (National Research Council) (1993). Nutrient requirements of fish. Washington D.C., Academy Press.

Olsen YA, Einarsdottir IE, Nissen KJ (1995) Metomidate anaesthesia in Atlantic salmon, *Salmo salar*, prevents plasma cortisol increase during stress. Aquaculture 134:155–168.

Owen JM, Adron JW, Middleton C, Cowey CB (1975) Elongation and desaturation of dietary fatty acids in turbot (*Scophthalmus maximus*) and rainbow trout (*Salmo gairdneri*). Lipids, 10, 528-531.

Park MR, Chang YJ, Kang DY (1999) Physiological response of the cultured olive flounder (*Paralichthys olivaceus*) to the sharp changes of water temperature. Journal of Aquaculture 12:221–228.

Park SH, Kim TW, Kim SB (2009) Characterization of physicochemical properties of collagen from shark (*Isurus oxyrinchus*) skin. Korean Journal of Fisheries and Aquatic Science 42:574–579.

Park SK, Kim MJ (2007) Effects of changing age structure of population on seafood consumption. Ocean Policy Research, 23, 1-26.

Park YH, Jang DS, Kim SB (1997) Processing of the Sea Food. Hyungsul Press, Seoul, Korea, pp 166-168.

Parry RM, Chandau RC, Shahani RM (1965) A rapid and sensitive assay of muramidase. Experimental Biology and Medicine 119:384–386.

Pham AM, Lee KJ, Lim SJ, Lee BJ, Kim SS, Park YJ, Lee SM (2005) Fish meal replacement by cottonseed and soybean meal in diets for juvenile olive flounder, *Paralichthys olivaceus*. Journal of Aquaculture 18:215–221.

Post G (1983) Nutrition and nutritional diseases of fish. In: Textbook of fish health. TFH Publications. Inc. Ltd., Neptune, NJ, USA, pp 199–207.

Ra KS, Suh HJ, Chung SJ, Son JY (1997) Antioxidant activity of solvent extract from onion skin. Korean Journal of Food Science and Technology 29:595–600.

Rathbone CK, Babbitt JK, Dong FM, Hardy RW (2001) Performance of juvenile coho salmon *Oncorhynchus kisutch* fed diets containing meals from fish wastes, deboned fish wastes, or skin-and-bone by-product as the protein ingredient. Journal of the World Aquaculture Society, 32, 21-29.

Rawles SD, Riche M, Gaylord TG, Webb J, Freeman DW, Davis M (2006) Evaluation of poultry by-product meal in commercial diets for hybrid striped bass (*Morone chrysops* ♀ × *Morone saxatilis* ♂) in recirculated tank production. Aquaculture 259:377–389.

Rena KJ, Hasan MR (2009) Impact of rising feed ingredient prices on aqua feeds and aquaculture production, FAO Fisheries and Aquaculture Technical Paper., pp 541.

Robert CR, Braden SL, Laprarie RJ (1993) Substitution of soybean protein for fish protein in formulated diets for red swamp crawfish *Procambarus clarkii*. Journal of World Aquaculture Society 24:329–338.

Rogie A, Skinner ER (1985) The roles of the intestine and liver in the biosynthesis of plasma lipoproteins in the rainbow trout, *Salmo gairdneri* Richandson. Comparative Biochemistry and Physiology Part B: Comparative Biochemistry 81:285–289.

Ronnestad I, Fyhn HJ (1993) Metabolic aspects of free amino acids in developing marine fish eggs and larvae. Reviews in Fisheries Science 1: 239-259.

Saglio PH, Fauconneau B, Blanc JM (1990) Orientation of carp *Cyprinm carpio* L to free amino acids from Tubifex extract in an olfactometer. Journal of Fish Biology 37:887-898.

Sandel LJ, Daniel JC (1988) Effect of ascorbic acid on collagen mRNA levels in short term chondrocyte cultures. Connective Tissue Research 17:11–22.

Sato T, Kikuchi K (1997) Meat meal as a protein source in the diet of juvenile Japanese flounder. Fisheries Science 63:877–880.

Schreck CB (1982) Stress and rearing of salmonids. Aquaculture 28:241–249.

Schrieber R, Seybold U (1993) Gelatin production, the six steps to maximum safety. Developments in Biological Standardization 80:195–198.

Shapawi R, Ng WK, Mustafa S (2007) Replacement of fish meal with poultry by-product meal in diets formulated for the humpback grouper, *Cromileptes altivelis*. Aquaculture 273:118–126.

Sivaloganathan B, Walford J, Ip YK, Lam TJ (1998) Free amino acids and energy methabolism in eggs and larvae of seabass *Lstes calcarifer*. Marine Biology 131: 695-702.

Steffens W (1994) Replacing fish meal with poultry by-product meal in diets for rainbow trout, *Oncorhynchus mykiss*. Aquaculture 124:27–34.

Sverdrup A, Kjellsby E, Krüger PG, Fløysand R, Knudsen FR, Enger PS, Serck-Hanssen G, Helle KB (1994) Effects of experimental seismic shock on vasoactivity of arteries, integrity of the vascular endothelium and on primary stress hormones of the Atlantic salmon. Journal of Fish Biology 45:973–995.

Toppe J, Aksnes A, Hope B, Albrektsen S (2006) Inclusion of fish bone and crab by-products in diets for Atlantic cod, *Gadus morhua*. Aquaculture 253:636–645.

Uyan O, Koshio S, Teshima S, Ishikawa M, Thu M, Alam MS, Michael FR (2006) Growth and phosphorus loading by partially replacing fishmeal with tuna muscle by-product powder in the diet of juvenile Japanese flounder, *Paralichthys olivaceus*. Aquaculture 257:437–445.

Verlhac V, Gabauda J (1997) The effect of vitamin C on fish health. Roche, Basel, Switzerland, pp 30.

Verlhac V, Gabaudan J (1994) Influence of vitamin C on the immune system of salmonids. Aquaculture 25:21–36.

Walrand S, Chiotelli E, Noirt F, Mwewa S, Lasse T (2008) Consumption of a functional fermented milk containing collagen hydrolysate improves the concentration of collagen-specific amino acids in plasma. Journal of Agricultural and Food Chemistry 56:7790–7795.

Watanabe T (1991) Past and present approaches to aquaculture waste management in Japan. In C. B. Cowey, & C. Y. Cho (Eds.). Nutritional strategies & aquaculture wastes (pp. 137-154). Ontario: Fish Nutrition Research Laboratory.

Wedemeyer GA, McLeay DJ (1981) Methods for determining the tolerance of fishes to environmental stressors. In: Pickering AD (ed) Stress and fish. Academic Press, London, UK, pp 247–275.

Wedemeyer GA, Yasutake WT (1977) Clinical methods for the assessment of the effects of environmental stress in fish health. US Fish and Wildlife Service 89:18.

Wendelaar Bonga SE (1997) The stress response in fish. Physiol Rev 77:591–625.

Wilson, R.P. (2002). Fish nutrition. (3rd ed.). California: Academic press.

Yoo SJ, Cho SM, Woo JW, Kim SH, Han YN, Ahn JR, Kim SY, Kim TW, Kim SB (2008) Processing and physicochemical properties of collagen from yellowfin tuna (*Thunnus albacares*) abdominal skin. Journal of Korean Fisheries Society 41:427–434.

Yoshinaka R (1989) Collagen of fish. In: Arai K (ed) Comparative biochemistry of muscular protein in aquatic animals. Kouseisha-Kosekaku, Tokyo, Japan, pp 81–90.

Yoshinaka R, Sato M, Ikeda S (1978) Distribution of collagenase in the digestive organs of some teleosts. Bulletin of Japanese Society of Scientific Fisheries 44:263–267.

Yoshinaka, R., Sato, M., Sato, S., Itoh, Y., Hujita, M., & Ikeda. S. (1985). Constituent proteins of muscle stromata from carp and Japanese mackerel. Fisheries Science, 51, 1163-1168.

Yoo, S.J., Cho, S.M., Woo, J.W., Kim, S.H., Han, Y.N., Ahn, J.R., Kim, S.Y., Kim, T.W., & Kim, S.B. (2008). Processing and physicochemical properties of collagen from yellowfin tuna (*Thunnus albacares*) abdominal skin. Journal of Korean Fisheries Society, 41, 427-434.

Zeitler MH, Kirchgessner M, Schwarz FJ (1984) Effects of different protein and energy supplies on carcass composition of carp (*Cyprinus carpio* L.). Aquaculture 36:37–48.

Zoccarato I, Gasco L, Sicuro B, Palmegia-No GB, Boccignone M, Bianchini ML, Luzzana U (1996) Use of a by-product from poultry slaughtering in rainbow trout (*Onchrhynchus mykiss*) feeding. Rivista Italiana di Aquacoltura 31:127–134.

2010 Fishery Production Statistics Yearbook, South Korea